Study Guide: Fundamentals of Industrial Hygiene

Third Edition

J. Thomas Pierce, Ph.D., CIH

Technical Advisers: George S. Benjamin, MD, FACS, FACOM
Jill Niland, MPH, CIH, CSP
Project Editor: Jodey B. Schonfeld
Interior Design and Composition: Impressions, Inc.

©1989 by the National Safety Council
All rights reserved.
Printed in the United States of America.
96 95 94   9 8 7 6

National Safety Council
Study Guide: Fundamentals of Industrial Hygiene
International Standard Book Number 0-87912-141-6
1M394

Product Number: 151.36-0000

Contents

Part Six—Industrial Hygiene Programs

Part Seven—Governmental Regulations and Their Impact

Answers

(There are no specific answers for questions on Chapters 24, 25, 27, 28, 29, and 31.)

Preface

Literally a generation of health and safety professionals, industrial hygienists included, have "grown up" with the *Fundamentals of Industrial Hygiene*. The ways in which they have learned the material it covers are as varied as the locations where they have studied.

A pervasive element of modern professional life is the need to constantly study and review technical material. The sheer volume of information that is appropriate for study dictates a means of highlighting important concepts and principles.

This *Study Guide* to the *Fundamentals of Industrial Hygiene* now enables us to focus in on the most important concepts of industrial hygiene. The third edition of the *Fundamentals* is organized with ease of comprehension in mind. It is divided into parts; each chapter contents page lists the main topics. Using the all new *Study Guide,* readers of the *Fundamentals of Industrial Hygiene* can quickly assess which areas need more work. This *Guide* offers creatively designed case studies, problems, and questions, which give its users a chance to apply the principles they have learned from the text. These exercises lead to an improved command of the subject; and will assist persons in acquiring skills in the field, setting up and running an industrial hygiene program; and successfully integrating an industrial hygiene program into the other areas of occupational safety and health.

During the past three years C. Lyle Cheever and I have labored hard to produce a continuing education series for practicing industrial hygienists which appears monthly in *Applied Industrial Hygiene.* The focus of the *Student Guide* is somewhat different, however. Here, we offer help to those learning the tenets of industrial hygiene for the first time—whether students or working professionals—this *Guide* will prove invaluable to those using the *Fundamentals of Industrial Hygiene.*

The background for these questions originated in many locations and with many of my own "teachers." However, as anyone who has ever taught knows, the greatest encouragement for learning comes from one's own students. The incentive to finish this project came from my students in Alabama and now in Virginia. It is dedicated to our friend Jon L. Barnett, CIH (1959–1986).

J. Thomas Pierce, Ph.D., CIH
Medical College of Virginia
Virginia Commonwealth University

1 Overview of Industrial Hygiene

1. Which one of the following professional groups might be expected to apply and interpret occupational health standards? (Name all that apply.)

   a. occupational physicians
   b. industrial toxicologists
   c. cardiologists
   d. industrial hygienists
   e. safety professionals

2. Which one one of the following acronyms represents the primary source for the original Permissible Exposure Limits incorporated in the Occupational Safety and Health Act Regulations?

   a. MAC
   b. MSHA
   c. ACGIH
   d. ANSI
   e. AIHA

3. Which one of the following methods is not recommended as an industrial hygiene corrective measure?

   a. substituting less harmful materials
   b. changing work processes
   c. relying upon the use of prophylactic drugs (e.g., EDTA for Pb exposures)
   d. housekeeping measures
   e. providing protective equipment

4. Classic ergonomic design studies would be least involved with:

   a. tool design.
   b. lifting.
   c. lighting.
   d. fume reduction.
   e. biomechanics.

5. Of the choices listed, which one is most significant in terms of entry of solvent vapors?

    a. inhalation
    b. ingestion
    c. subcutaneous injection
    d. eye

6. When comparing the terms toxicity and hazard, it is most appropriate to conclude that:

    a. they may have the same meaning.
    b. hazard is an intrinsic property of a material.
    c. toxicity depends upon the pattern of exposure.
    d. hazard depends upon the pattern of exposure.

7. For noise hazard purposes, damage risk criteria may depend upon: (Check all that apply.)

    a. total daily duration of exposure.
    b. frequency distribution of the sound.
    c. length of employment in the noise environment.
    d. sound energy.

8. A worker exhibiting signs, including a normal or low oral temperature, pallor, dizziness, and cool moist skin, is probably suffering from which one of the following disorders?

    a. heat syncope
    b. heatstroke
    c. heat exhaustion
    d. chilbain

9. Which one of the following categories of radiation is generally not considered to be an external hazard?

    a. alpha
    b. beta
    c. gamma
    d. neutron
    e. x-ray

10. A condition that usually affects workers operating tools, such as air hammers, air chisels, and chain saws, and which obstructs circulation of the hands is known as:

    a. byssinosis.
    b. bagassosis.
    c. Raynaud's syndrome.
    d. arthritis.

11. Terms that distinguish airborne matter (e.g., gas, fume, and vapor) may be said to:

    a. carry precise industrial hygiene meanings.
    b. be used interchangeably by industrial hygienists and the news media.
    c. have no particular relevance to industrial hygiene.
    d. refer only to environmental (ambient) pollutants.

12. Indicate all of the following choices that represent significant physiological routes of entry for industrial contaminants:

    a. accession
    b. inhalation
    c. dermal absorption
    d. ingestion

The following example shows the variety of stressors common in many industries. Industrial hygienists must anticipate, recognize, evaluate, and control aspects of such situations. One of the first steps in hazard assessment is to determine what kinds of stressors are involved.

13. Use your textbook to categorize the agents listed in this brief passage.

    A midwestern city's electric power plant uses shredded residential refuse and coal as a fuel for generation of electricity. Operating 24 hours a day, the plant has a total work force of 160 workers.
    During a survey, personal exposure monitoring was conducted for cadmium, total chromium, chromium VI (insoluble form), lead, nickel, respirable dust, respirable coal dust, and respirable crystalline silica. Bulk dust samples were analyzed to identify potential airborne contaminants. Area sampling was conducted for airborne microbial contamination using viable samplers. Heat stress was evaluated using a wet bulb globe temperature heat stress monitor.

14. Some of the agents listed in this passage may not be familiar to you, but as a trained professional you must be able to recognize significant threats to worker health. What basic strategies are used in the hazard recognition process?

_____

_____

_____

_____

_____

_____

_____

_____

_____

_____

_____

15. The passage indicates that personal exposure monitoring for a number of chemical contaminants was conducted. Using the results, what comparisons would be made against published recommendations, other criteria, or regulated concentrations?

_____

_____

16. Besides the survey information previously noted, what other key elements would be important in the evaluation phase?

_____

_____

② The Lungs

1. All of the following serve to condition and remove particulates except:

    a. turbinates.
    b. mucus.
    c. cilia.
    d. diaphragm.

2. Often an industrial hygienist must make an approximation concerning the amount of air that an individual might breathe during a work shift. Calculate the approximate total volume that a worker might breathe during such an eight-hour period.

3. Most of the oxygen transported in blood is:

    a. combined with hemoglobin in blood.
    b. bound to carboxyhemoglobin.
    c. bound to sulfmethemoglobin.
    d. free in solution.

4. The breathing mechanism is most closely regulated by:

    a. oxygen bound to hemoglobin.
    b. oxygen free in solution.
    c. carbon dioxide.
    d. none of the choices listed above.

5. Which of the following spirometry terms refer(s) to the volume of air which can be forcibly expelled during the first second of expiration?

   a. FVC
   b. FEV_1
   c. FEF_{25-75}
   d. FRC

6. Which one of the following most closely matches the correct meaning of the term *pneumoconiosis?*

   a. outpouring of liquid
   b. lung inflammation
   c. "dusty" lung
   d. friction between the visceral and parietal pleura

7. Exposure to cadmium oxide fume at various concentrations can cause which of the following?

   a. pulmonary edema
   b. renal tubular damage
   c. an emphysema-like condition
   d. all of the choices listed above

8. All of the following gases should affect the upper respiratory tract except:

   a. ammonia.
   b. hydrogen chloride.
   c. sulfur dioxide.
   d. ozone.

9. Considering the absorption of gases and vapors, which of the following represents the most important site for absorption?

   a. skin
   b. lung
   c. small intestine
   d. liver

10. After a normal expiration, the total lung volume remaining is called the functional residual capacity and is composed of:

    a. residual volume, tidal volume.
    b. residual volume, expiratory reserve volume.
    c. inspiratory reserve volume.
    d. tidal volume, inspiratory reserve volume.

11. Pulmonary function testing requires consistent and often fatiguing effort on the part of the test subject. If replicate (repeat) tests differ by more than ____?____ percent they are generally considered to be unsatisfactory.

12. The correct value for the partial pressure of oxygen at sea level and normal barometric pressure is most nearly:

    a. 160 mmHg.
    b. 760 mmHg.
    c. 600 mmHg.
    d.   76 mmHg.
    e. none of the choices listed above

13. Distinguish between external and *true* (or biochemical) respiration.

    _____

    _____

    _____

    _____

    _____

    _____

    _____

    _____

Case Study—Bakery Mixing Room

(Note: Case study questions 14–17 may require familiarity with material given in text Chapter 7 as well as text Chapter 2. You may wish to finish Chapter 7 before beginning this section.)

Seven workers started working in the mixing room of a bakery room only one year ago. None had previously worked in this trade. Since then, however, they have developed severe, fixed obstructive lung disease. Medical examination reveals a decreased FEV_1/FVC.

The bakery asked a team of industrial hygienists to investigate the causes. As part of the team, you conduct interviews and sampling and analyze materials used in the mixing room.

The mixing room is a large open room where three employees weigh and load a large variety of fragrances, flavorings, starch, and 50- to 100-lb bags of flour into one of three mixers. The loading and mixing tasks generate considerable dust. Area samples generally indicate a concentration of approximately 20 mg/m^3.

14. How would you evaluate the severity of the exposure from a health viewpoint?

_____

_____

_____

_____

_____

_____

_____

15. What defenses does the respiratory system have against the types of exposures described in the passage? A small group of the mixing room employees has developed severe, fixed obstructive lung disease within one year of first employment in this trade. Medical examination reveals a decreased forced expiratory volume to forced vital capacity ratio (FEV_1/FVC).

_____

_____

_____

_____

_____

_____

_____

16. Based upon reference information in text Chapter 7, within what general particle size range might these dusts appear?

_____

_____

_____

_____

_____

_____

17. What are the explosion characteristics of such dusts? What factors are used to evaluate the extent of explosion hazard for combustible dusts?

_____

_____

_____

_____

_____

_____

③ The Skin

Much of the material on human anatomy and physiology in this chapter may help you understand why occupational skin diseases occur in some individuals and not in others. This differential susceptibility and resistance can be tied to specific factors, whether these are genetic, environmental, or indirect.

1. In each of the following categories list an example appropriate to the factor in question. Reference to text Chapter 8 may be necessary to answer some questions.

   a. A genetic factor making the skin more susceptible to exposure to ultraviolet (UV) radiation:

   _____

   _____

   _____

   b. An environmental factor related to contact with photoreactive substances:

   _____

   _____

   _____

   c. Genetic factor affecting dissipation of heat:

   _____

   _____

   _____

2. Which of the following is a function of skin?

   a. temperature regulation
   b. protection against UV radiation
   c. barrier to microorganisms
   d. tactile sensation
   e. all of the above choices

3. The most common occupational disease appears to be:

   a. pneumoconiosis.
   b. asbestosis.
   c. fume fever.
   d. dermatosis.

4. Which of the following represents the primary barrier to absorption of materials via skin?

   a. melanin
   b. stratum corneum
   c. dermis
   d. stratum granulosum

5. Of the following traits, which one is the most characteristic of *sensitization dermatitis?*

   a. latency (induction) period
   b. irritation at the site of exposure
   c. secondary infection
   d. chemical burns

6. Coal-tar based chemicals can cause which of the following conditions?

   a. dermatitis
   b. folliculitis
   c. skin cancer
   d. all of the choices listed above

7. Which of the following is most important to regulating body temperature?

   a. apocrine sweat
   b. eccrine sweat
   c. sebaceous glands
   d. melanocytes

8. Briefly describe the process of dermal absorption of materials.

   _____

   _____

   _____

   _____

   _____

9. The hands of tradespeople are sometimes callused according to their occupations. In what general category of skin-related insult or injury does this fit?

_____

_____

_____

_____

10. Indicate two categories of dermal threats that can confront sewage and compost workers in an agricultural region. Even though skin might be affected by the agents in question, are other systems at risk as well? Explain.

_____

_____

_____

_____

11. From the information contained in Table 3-A in the text (p. 54), what chemical substances might have caused dermatitis on the feet and ankles of a railroad worker, who wears company-issued safety shoes without socks?

_____

_____

12. Why do percutaneous absorption rates through dorsal and ventral surfaces of the hand vary significantly?

_____

_____

4 The Ears

1. Which one of the following statements concerning hearing is *incorrect?*
   The ear:

   a. responds to frequencies in the general range of 20-20,000 Hz.
   b. responds to minute pressure fluctuations.
   c. responds over a broad range of intensities.
   d. canal possesses significant ability to protect the inner ear against high-level sound.

2. Which one of the following is not a component of the ossicular chain?

   a. malleus
   b. incus
   c. anvil
   d. stapes
   e. cochlea

3. In addition to transferring sound energy between the outer and inner ears, the middle ear serves:

   a. to amplify sound.
   b. to convert a pressure signal to an electrical signal.
   c. maintain vestibular balance.
   d. all of the choices listed above.
   e. none of the choices listed above.

4. In addition to occupational noise, which of the following conditions or disorders may be responsible for a hearing loss? (Check all that may apply.)

   a. rubella
   b. displacement of the ossicles
   c. earwax
   d. presbycusis

5. A sensorineural hearing loss is most often linked with an auditory decrement in which one of the following frequency regions?

   a. 4,000 Hz
   b. 4,000 cycles per minute
   c. 20 Hz
   d. 20,000 Hz
   e. 1 Hz

6. Which of the following factors may be important in the development of both temporary and permanent threshold shifts? (Check all that apply.)

   a. decibel level
   b. individual tolerance (genetic predisposition)
   c. frequency distribution of noise
   d. length of time in a noise-free environment

7. Which of the following phenomena represent important problems attributable to high level noise? (Check all that apply.)

   a. speech masking
   b. hearing damage
   c. sensorineural hearing loss
   d. annoyance and irritation

8. Sensorineural hearing loss may be caused by:

   a. excessive ear wax.
   b. high level noise.
   c. otosclerosis.
   d. all of the above.

9. The condition in which an individual senses sounds that are not of an external origin is called:

   a. speech misperception.
   b. paracelsus.
   c. tinnitus.
   d. presbycusis.

Case Study—Sound Levels in Shipyards

A new shipyard plasma arc welding operation is used to cut deck plates that are being recycled for other uses. The engineering literature on this process indicates that it involves very high temperatures; and that the plasma emanates from a mixture of electrons and gaseous ions formed when argon is forced through a high temperature arc.

Providing ventilation and respiratory protection has reduced air contaminants to acceptable concentrations but noise problems are severe. (In later chapters, you will better explore means to evaluate and measure noise; but here you are given information regarding basic noise problems and their effects on the shipyard workers.)

Measured sound pressure levels approximate 112–116 dBA in the immediate vicinity of the operation. Affected workers indicate that they have difficulty after work shifts understanding speech, and that their families complain regarding the audio settings necessary for them to hear radio and television broadcasts.

10. What well-known noise-related phenomenon is occurring among these workers following each work shift?

_____

_____

11. If acoustical treatments or other protective measures are not developed to control this noise hazard, what can be expected to occur in most workers exposed at these levels? If sound pressure levels were reduced to 85 dBA, what would be expected?

_____

_____

_____

_____

_____

_____

12. At what levels should workers be included in the hearing conservation program according to the provisions of the OSHA Hearing Conservation Amendment?

_____

_____

5 The Eyes

1. The aqueous humor occupies the region between the (?) and the (?).

2. The flow and pressure of the aqueous humor become problematic in terms of which of the following diseases?

    a. cataract
    b. keratitis
    c. otitis media
    d. glaucoma

3. The term *refraction* is important in understanding the process of transmission of light from the cornea to the focused retinal image. What is the meaning of the term *refraction*?

    _____

    _____

4. Which of the following eye tissues is/are primarily responsible for reducing light scattering within the eye?

    a. ciliary body
    b. choroid
    c. sclera
    d. retina

5. Which of the following might be considered the most significant defect for a crane operator?

    a. lack of binocular vision
    b. light compensation
    c. color vision
    d. correctable myopia

6. Which of the following is the principal remedy for a victim of an unidentified chemical splash incident? (The victim is still in the shop area.)

    a. boric acid solution
    b. commercial eye irrigants
    c. slightly alkaline mixture
    d. ordinary cool "tap" water

Case Study—Home Restoration

An industrial hygienist is contacted by a home restoration company that specializes in refurbishing older homes with plaster and lathe interior finishes. During the refurbishment process crews sand on existing plaster and refinish badly damaged areas with new plaster.

During this process, several workers have complained of eye irritation, particularly when sanding on the new plaster. Other problem areas involve preparing the paste using water, and mixing operations that involve bagged plaster.

7. Although the solution to at least part of this problem is fairly obvious, suggest suitable means of eye protection and indicate a guide to adequacy of safety eyewear.

8. A very good chemical explanation exists for the problem involving the new plaster and the apparent lack of problem with older plaster. The calcium hydroxide in plaster is eventually converted to the more physiologically inert calcium carbonate through the adsorption of carbon dioxide. (No answer is required here.)

9. Plaster is a mixture consisting of water, calcium hydroxide, and calcium sulfate (plaster of paris). From a chemical standpoint, which of these is likely to cause the greatest degree of damage?

10. Even if the old plaster was chemically inert with respect to the eye, what hazard potential would continue to exist? Hint: Remember that the eye risk may be due to both the chemical and mechanical properties of the plaster dust.

⑥ Solvents

1. Generalizations concerning the effects of organic (industrial) solvents generally conclude that:

   a. the lung will be affected.
   b. the liver will be affected.
   c. the kidney will be affected.
   d. a–c are all true.
   e. Such a generalization is not possible.

2. Which one of the choices shown below would likely represent the most effective industrial hygiene means of control for a solvent with a relatively high vapor pressure and a low TLV?

   a. general dilution ventilation
   b. local exhaust ventilation
   c. use of pedestal fans at strategic locations
   d. barrier cream application before the work shift
   e. none of the above choices

3. Which one of the following would not be used as a solvent? (Give a reason why it would not be used.)

   a. carbon disulfide
   b. dimethyl sulfoxide
   c. hydrogen sulfide
   d. benzene
   e. methyl-n-butyl ketone (2-hexanone)

4. In lower molecular weight aliphatic chlorine compounds, increasing the degree of chlorine substitution appears to:

   a. lower the TLV.
   b. raise the TLV.
   c. exert no effect on the TLV.

5. Which one of the following systems appears to be most susceptible to organic solvent exposures?

   a. renal
   b. central nervous
   c. lymphatic
   d. hematopoietic

6. Generally, it is presumed that if a flammable, toxic solvent is controlled within the prescribed limits of (?), then problems should not occur with regard to its (?).

   a. flammability, toxicity
   b. toxicity, flammability
   c. hazard, toxicity
   d. toxicity, hazard

7. Even though glycols have relatively low vapor pressures, which one of the following conditions would most increase human exposure risk?

   a. large uncovered vats at room temperature
   b. two similar glycols that are being mixed
   c. a high-pressure line carrying the glycol ruptures slightly, creating a mist
   d. parts being dipped in glycols

8. Which of the following solvents is/are generally regarded as the most toxic?

   a. methyl n-butyl ketone (2-hexanone)
   b. methyl isobutyl ketone
   c. methyl isoamyl ketone
   d. methyl isopropyl ketone

9. Effective control through local exhaust ventilation has the greatest possibility of limiting worker exposure by (?).

   a. inhalation.
   b. ingestion.
   c. skin contact.
   d. percutaneous injection.

10. In operations involving solvent spraying, the best protection for workers is afforded by placing primary and secondary reliance upon:

   a. respirators, general dilution ventilation.
   b. general dilution ventilation, respirators.
   c. local exhaust ventilation, respirators.
   d. administrative controls, respirators.

11. Neoprene gloves would likely offer the least protection against:

    a. machine oil.
    b. lubricating oils.
    c. hexane.
    d. benzene.

12. Measuring air concentrations of solvents provides little information concerning the likelihood of exposure via:

    a. inhalation.
    b. breathing.
    c. skin contact.
    d. nasopharyngeal contact.

13. When using portable direct-reading instruments for spot checks of organic solvent vapor concentrations, a presurvey check should consist of which one of the following? (Check all that apply.)

    a. battery check
    b. calibration against a known concentration of a vapor
    c. calibration against liquid concentrations of vapor
    d. zeroing using clean ambient air

14. (From text Chapter 6 "Addendum") When evaluating the degree of fire or explosion hazard from a solvent, which of the following should be considered? (Check all that apply.)

    a. upper explosive limit
    b. lower explosive limit
    c. fire point
    d. LC_{50}

15. (From text Chapter 6 "Addendum") The fire point typically relates to the flash point in which of the following ways?

    a. The fire point is lower.
    b. The fire point is higher.
    c. The the two values are identical.

16. (From text Chapter 6 "Addendum") Given a mixture of flammable liquids, which of their flash points is most critical?

_____

_____

17. (From text Chapter 6 "Addendum") Name at least three acceptable means of transferring flammable liquids.

    a. _____

    b. _____

    c. _____

Case Study—Auto Transmission Operation

Environmental and medical surveys were conducted to investigate employee exposures to oil mists and solvents and to evaluate suspected cancers occurring as a result of exposures at an automobile transmission facility. The investigation was requested by the employees' representative on behalf of approximately 50 employees. Analysis for oil mist in environmental samples indicated that exposures were below recommended levels. Analysis of solvent vapors indicated the presence of benzene and toluene at concentrations below recommended levels.

Medical surveillance revealed that five of 17 workers had skin problems related to contact with coolant oils. Though other employees reported similar skin problems, not all of their medical records were available; thus, no definite conclusion could be made about the presence of an occupationally-related cancer risk among the bar stock operators.

18. What chemical class is represented by the two compounds identified here? Suggest a suitable glove material for these two compounds and one unsuitable glove material.

    _____

    _____

    _____

    _____

19. What are the vapor/hazard ratios for toluene and ethylbenzene? Which compound would be preferred on this basis?

    _____

    _____

    _____

    _____

20. How can medical surveillance be conducted when a number of medical records for affected employees are not available? (No single correct answer exists for this question; you may wish to consult text Chapter 26, "The Occupational Physician.")

_____

_____

_____

_____

7 Particulates

1. In terms of hazard evaluation, which one of the following factors best distinguishes lead fume from lead dust?

   a. length of exposure
   b. concentrations
   c. size distribution
   d. none of the above choices

2. The current U.S. (NIOSH) convention for expressing asbestos levels in fibers/cc represents a number/volume concentration term. Check all other terms listed below that are consistent with this convention.

   a. mg/cubic meter
   b. fibers/cc
   c. fibers/mL
   d. million fibers/cubic meter

3. Of the listed forms of silicon dioxide, it is generally agreed that the most hazardous is:

   a. free crystalline silica.
   b. amorphous silica.
   c. silicates.
   d. No distinction is significant in terms of hazard.

4. Of the following factors, which one is least amenable to industrial hygiene measurement?

   a. composition of the dust
   b. percentage of free silica, if applicable
   c. particle size distribution
   d. predisposing genetic factors

5. Another term for Shaver's disease is:

   a. kaolinosis.
   b. anthrasilicosis.
   c. coal workers' pneumoconiosis.
   d. bauxite pneumoconiosis.

6. Which of the following factors would be important in the prevention of a dust explosion? (List all that apply.)

   a. presence of any ignition sources
   b. concentration of dust
   c. reasonable housekeeping measures
   d. supply of respirators

7. Which one of the following is not an accurate description of a characteristic of asbestosis?

   a. a malignant condition
   b. a recognized long latent period
   c. diffuse lung scarring
   d. interstitial lung scarring

8. Name four specific pneumoconioses and list a causative agent for each:

   a. _____

      _____

   b. _____

      _____

   c. _____

      _____

   d. _____

      _____

9. Historically, which one of the following mineral forms of asbestos has been used most often in the United States?

   a. chrysotile
   b. amosite
   c. crocidolite
   d. actinolite
   e. anthophyllite

10. How do the TLVs for the following crystalline forms of silica compare? (Please refer to the current ACGIH TLV/BEI Book for more information.)

    a. cristobalite
    b. quartz
    c. silica
    d. tridymite
    e. tripoli

11. Which is the most common naturally occurring form of free silica?

Case Study—Foundry Workers

Foundry workers being evaluated during a medical screening program had a higher-than-expected incidence of unequivocal pneumoconiosis (10/181 = 5.5%) and possible pneumoconiosis (6/181 = 3.3%). A review of previously collected sampling and analytical data indicates that foundry-related exposures to crystalline silica, heavy metals, coal tar pitch volatiles, methylene bisphenyl isocyanate, formaldehyde, and asbestos are generally within recommended standards. Of 81 personal samples collected for crystalline silica, 18 samples (one per individual worker) were above the recommended standard. Fifteen of these 18 workers were assigned to cleaning operations, most notably chipping and grinding.

12. Which one of the following compounds has not been implicated in the development of silicosis?

    a. tridymite
    b. cristobalite
    c. quartz
    d. bagasse

13. What device is most likely used in respirable dust sampling for silica?

---

14. Name three important steps in the preparation, calibration, and sample collection sequence for a respirable dust, such as silica.

    a. _____

    b. _____

    c. _____

    d. _____

    e. _____

⑧ Industrial Dermatoses

1. Of the following traits, which one is most characteristic of *sensitization* dermatitis?

    a. latency (induction) period
    b. contact irritation at the exact site of application
    c. secondary infections
    d. irritant response

2. Chloracne may be best described as:

    a. a form of folliculitis.
    b. adolescent skin phenomenon.
    c. chlorine-related disease.
    d. none of the choices listed above.

3. Of the following materials or methods, which one is probably the most effective in the control of industrial dermatitis?

    a. substitution for offending materials
    b. barrier creams
    c. local exhaust ventilation
    d. general dilution ventilation
    e. protective gloves

4. Which one of the following predisposing factors is most amenable to industrial hygiene control?

    a. atopic dermatitis
    b. age
    c. genetic characteristics of skin
    d. housekeeping within the work area
    e. seasonal effects

5. Which of the following would probably be the industrial hygienist's contribution to the diagnosis of an occupational skin disorder?

    a. detailed work exposure history
    b. ancillary diagnostic tests
    c. sites of involvement
    d. appearance of a lesion

6. Based upon considerations of lost workdays, the highest risk industry for occupational skin diseases is:

   a. poultry dressing.
   b. nonferrous foundry operations.
   c. small arms manufacture.
   d. paint formulation.

7. The skin notation in the ACGIH TLV/BEI Book indicates that:

   a. the chemical causes dermatitis.
   b. body burden due to skin absorption may be significant.
   c. the material may be of toxicological significance because of percutaneous absorption.
   d. none of the choices.

Case Study—Chemical Supply Company

Over a five-year period employees of a chemical (photography) supply company appeared to develop dermatitis through contact with triethylene glycol diacrylate (TDA). A multidisciplinary team toured their work area; talked with management, workers, and the plant physician; and reviewed records of these cases.

Dermatitis typically occurred on the hands and wrists of affected workers, occurring within 12–36 hours after initial exposure. Incidence was highest in chemical formulators; approximately 30% of these workers were affected. Maintenance personnel sporadically exposed to TDA were occasionally affected. Cases involving the forearm, hands, and wrists peaked during the months of May, June, and July.

8. From this brief description, does it appear these workers are suffering from irritant contact dermatitis or allergic contact dermatitis?

   _____

   _____

   _____

   _____

9. What significance should be attached to the finding that the dermatitis is more likely to involve the forearm area during the warm weather months of May, June, and July?

_____

_____

_____

_____

10. What key steps are involved in control of this dermatitis-producing hazard? Do these steps have to occur in any specific order?

_____

_____

_____

_____

Case Study—Duffel Bag Manufacturing Company

Dermatitis surfaced in a group of sewing machine operators and material handlers, who were manufacturing military duffel bags. Following medical interviews with these workers, samples of the cotton duck fabric used to make the bags were analyzed. Besides the cloth, the fabric contained 2,2'-methylene-bis(4-chlorophenol), synonym dichloroprene or G-4. The resulting condition appears to be a form of contact dermatitis.

There is not enough information available to permit a proper evaluation of the dermatitis-producing characteristics of this compound. Consequently, a toxicology research group will determine the toxic effects of these materials using animal tests.

11. What two major classes of skin-related problems due to contact dermatitis would you recommend that they evaluate?

_____

_____

12. A safety equipment vendor suggested that this problem might be controlled if the workers used barrier creams. What arguments might be mounted for and against this approach?

_____

_____

_____

_____

_____

_____

_____

13. The text provides four case examples of control: Epoxy spraying, machining, rubber manufacturing, and chemical manufacturing. Take any two of these examples, and compare and contrast the partial solution used. Are any generalizations evident from these experiences?

_____

⑨ Industrial Noise

1. Which one of the following estimates best describes the percentage of American manufacturing workers believed to be exposed to noise levels consistently above 90 dBA?
   a. 1%–10%
   b. 11%–20%
   c. 21%–50%
   d. >50%

2. Pitch or (?   ) refers to the number of cycles per second while intensity is better related to the (?   ) of sound.

   a. pressure, loudness
   b. frequency, loudness
   c. frequency, Hz
   d. Hz, frequency
   e. none of the choices listed above

3. The term used to describe the bending of sound around obstacles is:

   a. rarefaction.
   b. diffraction.
   c. compression.
   d. frequency.
   e. intensity.

4. Doubling of the actual sound pressure corresponds to a sound pressure level increase of (?   ) dB?

   a. 6
   b. 12
   c. 3
   d. 10

5. If the intensity of a sound source is doubled, the decibel level increase approximates which one of the following?

   a. 3
   b. 6
   c. 10
   d. 20

6. Which of the following conditions is most likely when the C-weighted reading is significantly higher than the A-weighted value?

   a. The sound has a large low-frequency component.
   b. The sound has a large high-frequency component.
   c. The sound energy is evenly distributed across the spectrum.
   d. Impulse noise predominates.

7. According to speech interference considerations, if the noise level approximates 90 dBA, what is the maximum distance in feet over which strained conversations can be conducted? (Consider strained conversations to approximate conditions of maximum sustained voice.)

   a. one
   b. five
   c. ten
   d. twenty

8. Which weighting network would give the least weight to a pure tone at 500 Hz?

   _____

   _____

9. Which of the following instruments would represent the method of choice for determining a worker's total noise exposure during his or her work shift?

   a. sound-level meter
   b. noise dosimeter
   c. audiometer
   d. octave-band analyzer

10. The usual relationship between the sound wave and the analyzer or recorded signal from a sound level meter is which one of the following?

    a. averaged
    b. peak
    c. RMS
    d. pulse

11. The preferred location for controlling a noise problem is at:

    a. the source.
    b. the worker's ear.
    c. the line of transmission.
    d. elsewhere.

12. Legal responsibility for providing a hearing conservation program resides with:

    a. employees.
    b. the employer.
    c. labor organizations.
    d. the industrial hygienist.

13. Industrial audiometry serves as:

    a. an alternative to sound surveys.
    b. an alternative to a hearing conservation program.
    c. an integral part of exposure dosimetry.
    d. an integral part of a hearing conservation program.

14. When conducting hearing protection surveys in the plants, the most important check is to examine muff (superaural) protectors for:

    a. hardening of seals.
    b. attenuation characteristics of the plastic bodies.
    c. metal fatigue.
    d. corrosion problems.

15. Superaural and insert-type protectors are least effective in attenuating (?) noise.

    a. high-frequency
    b. mid-frequency
    c. low-frequency
    d. 1,000–2,000 Hz

Case Study—Ceramic Tumbling Media Manufacture

A maker of ceramic tumbling media reports that potential hazards can exist with noisy conditions and dusts containing free silica and asbestos. As noted here, it is fairly typical for noise exposures to accompany chemical exposures.

(Most of the emphasis in this problem is on the noise aspects, but remember that industrial hygiene evaluation typically involves multiple stressors: even though calculations can focus on single agents, the entire work environment must be considered.)

Sound pressure level readings were made. Even though instantaneous readings of this nature can be quite different than the integrated results (obtainable through noise dosimetry), for this exercise, assume that the instantaneous readings are representative.

16. Determine whether the daily noise dose has been exceeded for an operator, based upon the following information:

Period Military clock (hr)	Sound Pressure Level (dBA)
1200–1300	84
1301–1600	84
1601–2000	89

17. Instantaneous readings were taken in the immediate vicinity of three selected pieces of equipment. Compute a combined sound pressure level, considering that the three devices operate at once.

Noise Source	Sound Pressure Level (dBA)
Abrasive chip boxing machine	89
Salvage machine	93
No. 5 pug mill	98

18. Using a knowledge of the noise control measures in the chapter, discuss the feasibility of noise reduction if no changes are made at the pug mill, while efforts are concentrated on the other two pieces of equipment. (For example, explain why noise control relative to other than the pug mill will not result in significant improvement.)

_____

_____

_____

_____

_____

_____

10 Ionizing Radiation

1. Which of the following materials would *not* be a good choice for the moderation of neutron energy?
   a. lead
   b. water
   c. paraffin
   d. cadmium

2. Protium, deuterium, and tritium are all isotopes of hydrogen, and each has one: (Check all that apply.)

   a. proton.
   b. neutron.
   c. electron.
   d. betatron.

3. Which of the following is a measure of activity for an isotope? (Check all that apply.)

   a. Rad
   b. Sievert
   c. Curie
   d. REM
   e. Becquerel

4. Alpha emitters should be classified as: (Check all that apply.)

   a. internal hazards.
   b. external hazards.
   c. forms of x-rays.
   d. forms of gamma rays.

5. Beta (negatron) emission changes the atomic number in which one of the following ways?

   a. increases by one
   b. decreases by one
   c. stays the same
   d. stays the same but mass number increases by two

6. Half-value layer figures vary depending upon the (?   ) in question and the (?   ) themselves.

   a. isotopes, calculations
   b. isotopes, shielding materials
   c. calculations, shielding materials
   d. radioisotopes, calculations

7. Once inhaled or ingested, radioisotopes will be absorbed, metabolized, and distributed:

   a. based upon their radioactive properties.
   b. in much the same manner as their nonradioactive counterparts.
   c. neither a. nor b.
   d. based upon the decay scheme involved.

8. The basic premise of occupational radiation protection dictates that:

   a. industrial workers represent an extremely select group of healthy individuals with resistance considerably different than that of the general population.
   b. if the dose rate is sufficiently low, body removal and recovery processes in cases of occupational exposures will probably be effective.
   c. any (occupational) exposure should not be tolerated.
   d. biological effects from artificial sources are trivial.

9. Accumulated occupational exposures can be expressed using which one of the following formulae (N = worker's age in years)?

   a. $5(N - 18)$
   b. $3(N - 18)$
   c. $18N$
   d. $3N$
   e. $5(N + 18)$

10. Which one of the following choices correctly applies to the use of distance as a means of radiation protection?

    a. Doubling the distance decreases exposure by a factor of two.
    b. Doubling the distance decreases exposure by a factor of four.
    c. Changing distance cannot be used to decrease exposure.
    d. Increasing the distance by a factor of four halves exposure.

11. High voltage equipment (e.g., equipment operating at greater than 15 kV) can be a significant source of what form of ionizing radiation? _____

Case Study—Chernobyl Nuclear Power Plant

On Saturday, 26 April 1986, unit 4 of the Soviet Chernobyl nuclear power plant exploded as the core suffered a prompt critical excursion. Other effects of this incident included death, injuries, and illnesses of workers at the immediate site, illnesses of residents in the vicinity of the plant, contamination and environmental effects throughout the region, and contamination and possible environmental effects throughout the ecoystem.

12. The principal long-term problem caused by such a nuclear reactor accident is contamination of the environment with Cs^{137}. What is the $t_{1/2}$ of Cs^{137}?

_____

_____

_____

13. If an inspector A received 25 REM in one hour at a distance of 25 m from the source, how much would inspector B receive at a distance of 40 m?

14. In the use of the formula

$$R/hr \text{ at } 1 \text{ ft} = 6 \text{ C·E·F}$$

when activity is expressed in millicuries and the calculation is performed without conversion, in what units should the dose rate be expressed? For what basic radiation type does this equation apply?

_____

_____

15. How do the risks from simply irradiated biological systems differ from the risks of contaminated biological systems?

_____

_____

_____

11 Nonionizing Radiation

1. Indicate at least two general consequences of biological damage associated with microwave exposures:

   a. _____

   b. _____

2. Choose from among the following ultraviolet regions, the one that represents the greatest risk to the eye (cataractogenesis). Consult text Figure 11–9, p. 235.

   a. 280–300 nm
   b. 300–340 nm
   c. 260–280 nm
   d. 240–260 nm

3. For an artificial device emitting ultraviolet (UV) light, describe the process that is used to obtain the equivalent energy density per 269.7 nm.

4. Name at least four quality determinants important in the provision of industrial lighting.

   a. _____

   b. _____

   c. _____

   d. _____

5. Consolidate named regions of the entire electromagnetic spectrum into eight identifiable groups. List them in the descending order of energy.

_____

_____

_____

_____

6. List a source and biological effect for each of the following types of radiation.

Radiation Type	Source	Biological Effect
(a) Ultraviolet		
(b) Infrared		
(c) Radiofrequency		

7. Microwave radiation is associated with all of the following applications except:

   a. radar.
   b. diathermy units.
   c. cooking.
   d. x-ray.
   e. communications equipment.

8. Energy per photon bears an inverse relationship with the wavelength of light; hence, it is important to be able to order named wavelength regions. Infrared, microwave, and radio waves all have wavelengths that are:

   a. longer than visible light.
   b. shorter than visible light.
   c. shorter than ultraviolet light.
   d. not comparable.

9. For health hazard evaluation, data concerning a device specifying the wavelength or frequency of nonionizing radiation would best be considered:

   a. complete, no supplementary information is necessary.
   b. incomplete, both wavelength and frequency must be specified.
   c. incomplete, intensity data are also necessary.
   d. incomplete, velocity information is needed.

10. Hazards associated with induction heaters are principally associated with which one of the following energy regions?

    a. UV
    b. radiofrequency
    c. infrared
    d. visible
    e. gamma rays

11. Which of the following sets of conditions is optimal for laser rooms?

    a. darkened room, general-purpose safety glasses
    b. darkened room, general-purpose safety goggles
    c. brightly-lit room, wavelength-specific goggles
    d. darkened room, wavelength-specific glasses

Case Study—Binder Manufacturing Plant

Six women employed in a facility manufacturing vinyl notebooks, and who use radiofrequency (RF) sealers in the manufacturing process, have been diagnosed with cancer of the breast. There is concern about the effects of levels of RF, particularly to women.

Approximately 300 workers at the facility are interviewed using a standard questionnaire. Industrial hygienists also take measurements of electric (E) and magnetic (H) fields at particular points along the workers' anatomy with respect to the machines. E- and H-field strengths appear to fall within both Occupational Safety and Health Administration standards and recommendations of the American Conference of Governmental Industrial Hygienists.

12. Look up the pertinent TLV, knowing that this device operates in the region of 30 MHz.

    _____

13. If this device was replaced by one operating at 300 MHz, what effect would this have on the TLV?

    _____

    _____

Case Study—Laser Laboratory

A scientific laboratory employs a range of lasers including Nd:YAG, copper-vapor, and xenon-fluoride types.

Type	Operating Wavelength(nm)	Operation	Pulse Repet. Frequency
Nd:YAG	106	Continuous	Single
Copper-vapor	510.6	Pulsed	6.08 KHz
Xenon-fluoride	351	Pulsed	<100 Hz

14. For the Nd:YAG laser, describe the basic wavelength region used and indicate the primary ocular structure at risk for persons exposed.

15. For the copper-vapor laser operating at 510.6 nm, laser-protective goggles with an optical density of 12 (at this frequency) are used. If the safety office indicates that adequate protection exists for those wearing the protective goggles, does a significant hazard exist for unprotected bystanders who may have contact with only the beam *reflection?* The laser's average output power is 400 mW, at a power density of 51 W/cm².

16. In addition to the data shown in the table, what other criteria would be used to determine occupational exposure limits for application to a wide variety of working environments?

17. Space limitations force the location of all three lasers in a single laboratory space. In light of protective eyewear required, will it ever be possible to operate more than one unit at a time?

12 Temperature Extremes

1. In the net heat exchange equation, which terms are always positive? Which terms are always negative?

    _____

    _____

2. What is the maximum core temperature that can be recommended for workers?

    _____

3. Which one of the following drugs can significantly affect heat acclimatization of workers?

    a. sedatives
    b. diuretics
    c. antispasmodics
    d. all of the choices listed above

4. A heat-related illness in which the patient has hot, dry skin, a core temperature of 41 degrees C, and convulsions would best be classified as:

    a. heat syncope.
    b. heat exhaustion.
    c. heatstroke.
    d. heat cramps.

5. A heat-related illness in which the patient has moist, clammy skin, low blood pressure, and a thready pulse, a normal oral temperature, but elevated rectal temperature, would best be classified as:

    a. heat exposure.
    b. heat exhaustion.
    c. heatstroke.
    d. heat cramps.

6. Which one of the following is the single, best means for immediate treatment of a heatstroke victim?

   a. reduce body temperature
   b. infuse fluids
   c. maintain airways
   d. restore breathing

7. Which mode of heat exchange between an individual and the environment is a function of air speed and the difference in vapor pressure between perspiration on the skin and in the air?

   a. convection
   b. radiation
   c. conduction
   d. evaporation

8. Which one of the following conditions best represents the result of prolonged reduction of core temperature?

   a. hypoplasia
   b. hypostasis
   c. hypothermia
   d. hypotaxis

9. How do heat acclimatization regimens differ for new versus previously acclimatized workers?

   _____

   _____

   _____

   _____

10. The difference between the natural wet bulb temperature and the dry bulb reading given a constant windspeed, can be used to provide a measure of:

    a. radiant heat.
    b. dry bulb temperature.
    c. wet bulb temperature.
    d. relative humidity.
    e. saturation.

11. The use of an aluminum screen placed in the transmission path of irradiance from glass melting ovens would most effectively reduce which one of the following heat components?

    a. radiant energy
    b. humidity
    c. air temperature
    d. wet bulb temperature

Case Study—Electric Power Plant

Workers in a municipal electric power plant are experiencing heat stress problems and may also be overexposed to various particulate materials. This facility uses shredded residential refuse and coal as a fuel to generate electricity.

Despite the frequency of complaints, it is difficult to monitor extended activities in the hot work area at the actual time of the industrial hygiene surveys. Consequently, the measurements obtained were heat stress readings taken in areas where work, primarily maintenance operations, might be performed.

Wet Bulb Globe Temperature (WBGT) heat stress readings were recorded at 16 C (61 F) to 29 C (84 F) during the March survey and 32 C (90 F) to 39 C (102 F) during August.

12. What principal forms can heat-related stress take in terms of illnesses or injuries?

    _____

    _____

13. Using Figure 1 for Permissible Heat Exposure Threshold Limit Values (current ACGIH TLV/BEI Book), determine the appropriate work-rest regimen for worst-case March exposures if the maintenance activities are estimated to require a work rate of 400 kcal/hr?

    _____

14. A plant engineer suggests that it would be easier to install numerous WBGT indicators throughout the station, than to continue making WBGT measurements. What is the principal limitation in the use of this instrument?

    _____

    _____

15. Which one of the following could individually offer a *significant* solution to problems occurring during August?

    a. work-rest regimen
    b. water-cooled garments
    c. use of electrolyte replacement beverages
    d. enhancing tolerance to heat

Case Study—Iron Foundry Workers

During hot summer months, floor personnel in an iron foundry have an increased incidence of heat-related injury. Corporate industrial hygiene policy dictates that both WBGT and HSI measurements must be made.

While the primary application of the WBGT data is in terms of medical surveillance and administrative controls, the HSI data are forwarded to the plant engineering department where they are used to evaluate engineering controls.

16. From the data provided here perform calculations necessary to arrive at the HSI and WBGT indices.

$$t_g = 100 \qquad \text{Air speed} = 500 \text{ ft/min}$$
$$t_{db} = 93 \qquad \text{Metabolism} = 1100 \text{ Btu/hr}$$
$$t_{wb} = 89 \qquad \text{Vapor pressure} = 17 \text{ mmHg}$$
$$M + R + C = 1200 \text{ Btu/hr}$$

17. In the example, what problem would result from the estimation of heat stress using the effective temperature index?

    _____

    _____

    _____

    _____

13 Ergonomics

1. Which of the following positions is associated with the lowest measure of internal compression forces?

   a. leaning backwards in a chair with a small lumbar backrest, arms supported
   b. sitting upright in a chair with a small lumbar backrest, rms hanging
   c. sitting slightly forward in a chair with a small lumbar backrest, arms hanging

2. Estimate the forces exerted in the lumbar spine when standing. Seated with arms hanging.

   _____

   _____

3. The NIOSH Work Practices Guide for Manual Lifting allows characterization of work processes involving:

   a. lowering.
   b. lifting and pushing
   c. carrying and lifting.
   d. lifting.

4. A single fixed seat height of 41 cm (16 in.) corresponding to the 50th percentile popliteal height of a mixed male-female population accommodates which one of the following groups?

   a. 95% of the male population
   b. 95% of a mixed population
   c. a relatively minor proportion of the mixed population
   d. no population segment at all

5. Which one of the following represents the best definition of ergonomics?

   a. the study of energy
   b. the field devoted to minimizing biological damage from ionizing radiation
   c. the study of human interactions with the work environment
   d. computer-assisted design (CAD)

6. Any action of a workplace-related external vector upon the human body, whether physiological or psychological, can best be classified as:

    a. work stress.
    b. work strain.
    c. ergonomic syndrome.
    d. none of the choices listed above

7. The most sensitive index suitable for monitoring physiological work is:

    a. heart rate.
    b. oxygen consumption.
    c. respiratory minute ventilation.
    d. heart stroke volume.

8. In an automated human-machine system, a human acts as:

    a. power source and controller.
    b. controller only.
    c. power source only.
    d. monitor.

9. Which of the following is likely the most important sensory input for human decision-making?

    a. visual
    b. auditory
    c. vibratory
    d. olfactory

10. Evaluate the validity of this statement:

    If machines are designed to optimize mechanical efficiency, human comfort automatically follows.

    _____

    _____

    _____

11. The best use of anthropometric data will permit the designer to plan for equipment that accommodates:

    a. only the average person (at the 50th percentile).
    b. 5th–95th percentile range.
    c. 25th–75th percentile range.

Case Study—Communications Workers

A group of communications workers reported an excessively high frequency of chest pain. After researchers analyzed epidemiological data, it appeared the chest pain occurred twice as frequently among VDT users than among nonusers, and the reports of chest pain increased with the amount of VDT use. The rate of reported chest pain also increased as the employees' perception of their ability to control their work situation decreased.

12. What stress-strain combinations could cause or possibly affect the symptoms described here? (Hint: The stress represents the external agent while the strain represents the physiological response.)

_____

_____

_____

_____

_____

_____

_____

13. Use material from Chapter 11, "Nonionizing Radiation," and this chapter to indicate important factors associated with physiological strain in VDT users.

_____

_____

_____

_____

_____

_____

_____

14. Which of the following is normally not listed as an adjustable feature of a video workstation:

    a. seat height
    b. table position
    c. focal length
    d. monitor position
    e. footrest position

Case Study—Commercial Cleaning Facility

Cumulative trauma disorders occurred in workers at a commercial uniform cleaning facility. Worker complaints centered around aching, numbness, clumsiness, and swelling of the hands and wrists.

Ergonomic evaluation of the jobs or work tasks suspected to be associated with cumulative trauma disorders consisted primarily of documentation of the hand/arm postures during exertion, the respective forces necessary, and the frequency. The analysis was conducted through the preparation and analysis of videotapes and photographs.

The analysts categorized the tasks as low-risk (15%), medium-risk (75%), and high-risk (10%). Further analysis of high-risk jobs indicated the tasks required repetitive hand and wrist movements and medium amounts of muscular force.

(Questions 15 and 16 require rather general answers.)

15. What other assessment strategies might have been used?

16. What control strategies could be employed?

14 Biological Hazards

1. Evaluate whether or not each of the following factors could be a major contributor to the occurrence of infectious diseases among workers in a biological research laboratory. Justify each choice:

    a. poor working habits: _____

    _____

    _____

    b. inadequate training: _____

    _____

    _____

    _____

    c. aerosol formation during routine laboratory procedures: _____

    _____

    _____

    _____

    d. improper handling of laboratory animals: _____

    _____

    _____

    _____

2. For each of the following infectious or parasitic diseases, evaluate whether or not occupational contact is feasible. If you answer yes, then list at least one occupation that is potentially at risk:

   a. Brucellosis: _____

   _____

   _____

   _____

   b. ORF: _____

   _____

   _____

   _____

   c. Histoplasmosis: _____

   _____

   _____

   _____

   d. Leptospirosis (Weil's disease): _____

   _____

   _____

   _____

   e. Hepatitis B: _____

   _____

   _____

   _____

3. For each of the following occupational groups, list an infectious or parasitic disease that constitutes a potential risk for the group:

    a. Biomedical repair technicians calibrating and repairing perfusion apparatus:

    _____

    _____

    b. Industrial hygienists investigating the causes of an outbreak of "humidifier fever":

    _____

    _____

    c. Dentists and dental hygienists performing restorations in an older dental office containing draperies and porous floor coverings:

    _____

    _____

    d. Slaughterhouse workers handling carcasses, hides, other animal remains:

    _____

    _____

4. The agent responsible for legionnaires' disease belongs in which one of the following categories?

    a. fungi
    b. bacteria
    c. viruses
    d. rickettsiae

5. Anthrax would typically be associated with which one of the following occupational groups?

    a. barbers
    b. veterinarians
    c. beauticians
    d. dentists

6. The agent responsible for Rocky Mountain spotted fever belongs in which of the following categories?

    a. fungi
    b. bacteria
    c. viruses
    d. rickettsiae

Case Study—Research Facility

An industrial hygienist is called to investigate the cause of a sudden outbreak of upper respiratory illness in a 15-story, academic research building housing animal facilities, research laboratories, and offices.

An examination of the aging roof-mounted air conditioning system reveals an absence of preventive maintenance. Stagnant water was observed to stand in the drip pans mounted below these units. On the third floor of the building, a laboratory "flood" had occurred that resulted in soaking of sound liner inside the duct work. Several workers in the animal handling facility reported symptoms preceding this incident by three weeks.

Sampling was conducted on several floors using a professional organization's bioaerosol committee protocol, employing sieve impactors operating at flow rates of 1 cfm. Viable fungi and actinomycetes were collected onto malt extract plates, while bacteria were cultured on soy peptone agar.

7. Explain what role "zoonoses" could play in such an incident.

    _____

    _____

    _____

    _____

    _____

    _____

8. What classes of organisms could be responsible if the diagnosis indicated hypersensitivity pneumonitis?

    _____

    _____

9. What significance would you attach to the poor maintenance of the air handling system and/or the laboratory flood?

_____

_____

_____

_____

_____

_____

Case Study—Biohazards

A clinical reference laboratory is undertaking a new project to lyophilize (freeze-dry) blood and urine samples that will later be analyzed for metals content. As part of the biohazard control program, biological safety cabinets are being ordered to contain the lyophilization apparatus. Since the latter analysis is strictly chemical in nature, no cultures are involved, hence contamination concerns are minimal.

10. From the information presented here would Class I or Class II cabinets be necessary? Explain.

_____

_____

_____

11. Workers in this laboratory have previously used only laboratory (chemical) fume hoods for transfers of toxic and flammable liquids and the new cabinets do not appear strikingly different. The laboratory safety officer has insisted upon a labeling effort directed at all hoods in the laboratory. What do you think justifies his concern?

_____

_____

_____

_____

15 Industrial Toxicology

Case Study—Research Compound

An industrial hygienist reviews information about a compound with characteristics listed in the following paragraphs. The compound has recently been synthesized in the research laboratory. It is necessary for the industrial hygienist to develop an "in house" occupational exposure limit and accompanying documentation. Use information provided in text Chapter 15, "Toxicology," to interpret the information in this passage.

Exposure to this compound produces nasal and bronchial irritation and liver and kidney impairment in animals; the substance readily penetrates the skin and is highly irritating to the eyes. The single oral LD_{50} in rats is 1.05 g/kg and the skin LD_{50} for 24-hour contact is 0.5 mg/kg.

A one-hour exposure to the concentrated vapor was not fatal in rats, nor was an eight-hour exposure to 8,000 ppm. Rats were exposed for eight hours daily to a concentration of 18,000 ppm (for a total of five days); after the first day, all animals showed severely reddened nasal mucosa or had respiratory distress; and there was one fatality (kidney and liver congestion).

Somewhat uncontrolled human exposures at a concentration of 12,000 ppm resulted in a subject reporting nose irritation after one minute and coughing after 90 seconds. Skin contact poses a moderately high degree of hazard, which diminishes as the product is diluted with water to less than 25%.

1. How would the health effects that are listed here be organized within the context of the classifications shown in text Table 15-G, p. 377?

    _____

    _____

    _____

    _____

2. What is the primary physiological response? _____

3. Would the animal data appear to apply to humans? Why or why not?

_____

_____

_____

4. Based upon the information presented, do you think it would it be necessary to develop occupational exposure limits for chronic, relatively low-level exposures as well as acute, potentially life-threatening incidents? Explain.

_____

_____

_____

Case Study—Inhalation Study

This fictional report illustrates important principles of using animal data to evaluate risk in the occupational (human) population. The reader should apply concepts covered in text Chapter 15, "Toxicology," to actual data.

The final report on the results of an inhalation study of a gaseous compound was released. This was a two-year inhalation study in which three groups of male and female Fischer 344 rats (120 rats/sex/group) were exposed at concentrations of 100, 33, and 10 ppm, for six hours per day, five days per week. Two groups of rats were exposed to air only and served as controls.

Based upon histological (tissue) evaluation, the researchers concluded that the incidence of mononuclear cell leukemia and of peritoneal (chest cavity) mesothelioma was significantly increased as a result of the exposures. The incidence of mononuclear cell leukemia in female rats was dose-related, increasing with exposure concentration.

A statistically significant increase in mononuclear cell leukemia was observed only in the group of female rats exposed at 100 ppm. For females exposed at 33 and 10 ppm, the cumulative percentage of animals developing leukemia was significantly higher than that for one control group, and for both control groups combined, but was not higher than the incidence for the second control group.

5. What is meant by the term *dose response* as it is used here or in studies of a similar nature?

_____

_____

_____

_____

_____

_____

_____

_____

6. What category of study is this with respect to the total length of the exposures? Is the length of the study with respect to animal lifetimes consistent with likely human working history exposures?

_____

_____

_____

_____

_____

_____

_____

7. Comment upon the usefulness of these data in formulating a TLV; what other kinds of information would be desirable in developing a TLV?

_____

_____

_____

_____

_____

_____

(The following are not case study questions.)

8. In order for an LD_{50} to convey useful information, which of the following must be specified? (Check all that apply.)

   a. species
   b. route of administration
   c. flash point
   d. time period of observation

9. A table showing the LD_{50}s for a number of materials demonstrates:

   a. mortality.
   b. chemical safety.
   c. hazard potential.
   d. morbidity.

10. Which of the following factors is considered least important in evaluating exposure to toxicant materials?

    a. weaning age of the animals
    b. route of administration
    c. duration of the exposure
    d. frequency of the exposure

11. The use of a TLV (as a time-weighted average) is least applicable in which of the following situations?

    a. chronic liver toxicity
    b. pneumoconioses
    c. narcosis
    d. allergic sensitivity

16 Evaluation

1. Which of the following statement(s) is/are true concerning TLVs? (Check all that apply.)

   a. (for irritants) A small percentage of workers may experience discomfort even when exposures are within current TLVs.
   b. A smaller percentage of workers may be affected more seriously by aggravation of existing conditions.
   c. A direct comparison can be made between TLVs and LD_{50}s.
   d. A direct comparison can be made between TLVs and PELs.

2. Which one of the following is correct?

   a. Occupational aspects of environmental health are designated as industrial hygiene.
   b. Environmental aspects of occupational health are designated as industrial hygiene.
   c. Occupational aspects of industrial hygiene are designated as environmental health.
   d. Occupational and environmental aspects of health are known as industrial hygiene.

3. Which one of the following provides the best example of the rather long latency period necessary for the production of occupational disease?

   a. allergic responses
   b. histamine release
   c. bronchospasm
   d. lung fibrosis

4. Which one of the following best represents the link between exposure by both respiratory and dermal routes and absorbed dose?

   a. general air monitoring
   b. personal monitoring
   c. biological monitoring
   d. medical surveillance for health effects

5. Which one of the following types of samples is not addressed in any of the current ACGIH BEIs?

    a. saliva
    b. mixed-exhaled air
    c. end-exhaled air
    d. urine
    e. red blood cells

6. Which one of the following meets the time limit definition appropriate to a STEL value?

    a. 10 minutes
    b. 15 minutes
    c. 20 minutes
    d. 30 minutes
    e. 45 minutes

7. Which one of the following choices is the oldest set of recommendations for occupational exposure criteria that are regularly updated?

    a. Hygienic Guides
    b. Threshold Limit Values
    c. NIOSH Criteria Documents
    d. Permissible Exposure Limits (OSHA)

Case Study—Chemical Manufacturing and Packaging

Exposures in a plant manufacturing and packaging various institutional and janitorial maintenance chemicals and allied products were evaluated. Workers reported symptoms, including headache, nasal irritation, muscular weakness, and nonspecific respiratory problems.

Realizing nearly 1,600 active raw materials were in use, investigators conducted industrial hygiene surveys at the plant. Interpreting results was complicated because production runs were not frequently coordinated with the surveys. Additionally, the number of chemicals present was staggering.

Informal medical interviews conducted with nine workers indicated a variety of symptoms that varied greatly in severity and frequency. Symptoms consistent with exposure to organic solvents and other contaminants were reported, including peripheral weakness and irritation of the nose, throat, and respiratory tract.

8. If continuous monitoring in the form of general area sampling was used, what problems might be averted? What deficiency would still exist in terms of estimating individual worker's exposures?

    _____

    _____

    _____

    _____

    _____

9. Given the obvious shortfalls indicated here, how might biological monitoring relate to air-sampling results?

    _____

    _____

    _____

    _____

    _____

    _____

17 Methods of Evaluation

1. The vapor pressure of a solvent (MW = 100) at 25 C and 1 atm pressure is 400 mmHg. Calculate the concentration resulting from a substantial spill in a small space that is left uncleaned. Express the answer in both ppm and mg/m^3.

2. How many mL of liquid CCl_4 must be evaporated into a 5-L bag at 25 C and 1 atm pressure, if a final concentration of 25 ppm (v/v) is desired? What fraction or multiple of the current TLV (time-weighted average) would this represent? (MW = 154; density = 1.5 g/mL)

3. Calculate the time-weighted average concentration given the following consecutive time periods sampled for a single worker:

Sample No.	Time Period (mil. clock)	Sampling Results (ppm)
1	0915–1030	320
2	1031–1210	250
3	1211–1600	350

4. In each of the following examples determine whether the sample illustrated better matches the characteristics of a breathing zone (personal) sample or environmental (area) sample.

   a. Samplers operate in a chemistry stockroom after an extended vacation, during which time the HVAC system was not operated. They are positioned on counters near the fume hood, the sink, and a reagent shelf;

   b. A worker in an animal feed extrusion room wears a passive monitor that allows measurement of both ethylene oxide and formaldehyde concentrations over an eight-hour period; _____

5. In making recommendations, it is often necessary to choose operations from among alternatives based upon incomplete health hazard information. Indicate which of the following cleaning operation(s) represent significant hazards.

   a. wet mopping
   b. use of compressed air
   c. use of vacuum cleaning devices
   d. steam cleaning

6. A good beginning approach to hazard recognition is to survey the literature regarding "typical" or expected concentrations associated with particular processes or operations. Name at least three operations given in this chapter, "Methods of Evaluation," for which information is listed regarding attendant chemical and physical hazards.

   a. _____

   b. _____

   c. _____

7. Outline basic strategies that are required when personal monitoring has indicated that exposures:

   a. exceed the action level but do not exceed the permissible exposure limit (See Glossary, Appendix F): _____

   _____

   _____

   _____

   _____

   b. exceed the permissible exposure limit: _____

   _____

   _____

   _____

   _____

   c. approximate 1/100 of the action level: _____

   _____

   _____

   _____

   _____

8. How many liters of air contaminated to the extent of 100 $\mu g/m^3$ must be collected if the chemist performing the analysis requires a minimum of 25 μg for the analysis? How long would this take when the air-sampler operates at 2.5 L/min?

9. Name three steps that can help to ensure quality control relative to industrial hygiene measurements:

   a. _____

   b. _____

   c. _____

10. Which of the following choices represents the single most important characteristic of evaluation of the worker's environment?

   a. Samples must represent the worker's daily TWA exposure.
   b. Samples must come from the first shift of the day.
   c. They must always be refrigerated before analysis.
   d. They should only be collected by a certified industrial hygienist.

   (Competent selection of an evaluation technique is not an issue.)

18 Air-Sampling Instruments

1. For each of the following collection devices, sketch or list the correct sequence of equipment necessary to sample chemical contaminants, starting with the air inlet and ending with the air mover. Collection devices must be oriented in a particular way; therefore, list the components in the correct order, as well.

   a. gas-wash bottle (midget impinger)
   b. mixed-fiber filter
   c. adsorbent tube

2. Which one of the following contaminants would not be a good candidate for sampling using silica gel tubes?

   a. methanol
   b. aniline
   c. aromatic amines
   d. silicates

3. Which one of the following is a primary standard for airflow calibration?

   a. rotameter
   b. dry-gas meter
   c. soap-bubble meter
   d. wet-test meter

4. Which one of the following is the source of contamination during transportation of samples?

   a. shipping charcoal tubes with bulk samples of solvent
   b. shipping MCE filter with bulk asbestos samples
   c. shipping MCE filter with insulation materials
   d. shipping charcoal tubes with passive samplers

5. Even though cyclone collectors are not particularly efficient for collecting small particles, they find application in air-sampling devices because they:

   a. effectively collect fine particles.
   b. pass fine particles on to the filter.
   c. have no effect upon collection of fine particles.
   d. resemble venturis.

6. The most important reason why an electrostatic precipitator is not considered a reasonable collection device for asbestos particles is because of the:

   a. unreasonable expense involved.
   b. lack of electrostatic charge on asbestos particles.
   c. plugging problems.
   d. power requirements of the device.

7. Indicators or detector tubes are sometimes considered a quick means for determining air concentrations of contaminants. Which one of the following is not an advantage of their use?

   a. simplicity of operation
   b. simplicity of data interpretation
   c. ability to collect grab samples
   d. convenience

Case Study—Steel Tubing Plant

A group of workers employed in a steel tubing plant requests an industrial hygiene evaluation of working conditions. They are concerned about the risk of cardiovascular disease that may be associated with their occupation. Medical evaluation confirms a significant excess prevalence of cardiovascular disease and a significant increased prevalence of respiratory symptoms among the workers.

Environmental samples are taken for total chromium, hexavalent chromium, iron oxide, particulate fluoride, manganese, lead, and welding fume.

8. Calibration for a sampling pump was done using a soap-bubble meter. Triplicate runs are shown below; indicate the flow-rate in mL/min, liters/min.

Trial	Time (sec)	Volume (mL)
A	22	720
B	23	750
C	21	680

9. If the tared weight of the filter used for manganese sample was 0.1280 g and its finished weight, 0.1994, calculate the concentration sampled if the pump indicated in question 8 operated for 400 min.

10. List at least three considerations that would influence the selection of sampling equipment suitable for the metals listed here?

a. _____

_____

_____

b. _____

_____

_____

c. _____

_____

_____

19 Direct-Reading Gas and Vapor Monitors

1. The electromagnetic spectrum was discussed in Chapter 11; that material is important now, since spectroscopic and colorimetric devices are used as direct-reading gas and vapor monitors. Indicate at least one device mentioned in Chapter 19, "Direct-Reading Gas and Vapor Monitors," that operates in each of the following regions:

    a. ultraviolet: _____

    b. visible: _____

    c. infrared: _____

2. In combustible gas indicators using the Wheatstone Bridge design, it is common to replace both filaments if one (typically the reactive filament) has broken. Why is it necessary to replace these with a perfectly matched pair, in terms of resistance?

    _____

    _____

    _____

    _____

    _____

(Questions 3–5 relate to this paragraph.)

A hypothetical gas has a lower explosive limit of 0.8% (v/v) and an upper explosive limit of 5.4%. An industrial hygienist performing a survey obtains a reading of 25%, as measured on the dial face.

3. Perform necessary calculations to indicate the actual concentration in ppm (v/v).

4. How does this problem illustrate the difficulty, if not impossibility, of using a combustible gas indicator to assess low concentrations, which are primarily significant in terms of health risk?

   _____

   _____

   _____

   _____

   _____

5. Why would alternatives to a combustible gas indicator be necessary to monitor flammable vapors in a space containing silicone vapors?

   _____

   _____

   _____

   _____

6. Indicator or detector tubes are sometimes considered as a quick means of determining concentrations of air contaminants. Which of the following does not represent an advantage attending their use?

    a. simplicity of use with respect to determining the sampled concentration
    b. simplicity of interpretation concerning worker exposures
    c. convenience
    d. analytical simplicity

Case Study—Distribution Warehouse

A large beer distribution facility asks an industrial hygienist to evaluate carbon monoxide (CO) exposures occurring inside its warehouse. During morning start-up, gasoline-powered delivery trucks are started inside the facility. Elsewhere in the building, gasoline-powered forklift trucks are used to unload railroad cars carrying beer kegs.

Carbon monoxide exposures occur when the beer trucks are started; they rapidly decline when the trucks leave around 0815 hours. In the delivery truck starting area, a solid electrolyte-type carbon monoxide monitor with a recorder readout is positioned in the center of the room and used to determine CO concentrations. Results from four area samples follow:

Sample	Concentration (ppm)	Time (hours)
A	220	0800–0815
B	45	0816–0830
C	30	0831–0930
D	17	0931–1600

7. Calculate a time-weighted average using these data. Realizing that this is an area sample, comment as to whether or not employees are being overexposed on either a short- or long-term basis.

Forklift truck operations occur on a more consistent basis and normally last 2–4 hours. A forklift truck operator from the keg loading area wore a personal monitor for carbon monoxide throughout the day. It indicated that levels of CO there reached 100 ppm 20 minutes after his forklift truck was used to stack beer barrels, and remained at this level until the procedure was completed.

It typically takes 3.5 hours for him to complete the procedure. An hour after the procedure was completed, the level of CO was still 60 ppm. Five hours later the concentration decreased to 7–13 ppm, which is the normal background level in this area.

8. Indicate whether the forklift truck operator has been overexposed to CO. It may be helpful to prepare a graph of concentration versus time.

9. What is the typical principle of operation associated with the monitoring devices mentioned in this problem?

10. Had monitoring been requested in a brewery, what contaminant would likely have been monitored? What technology exists to monitor this contaminant?

(Questions 11 and 12 are not related to the previous case study.)

11. Describe at least three principles of operation for oxygen monitors.

    a. _____

    _____

    b. _____

    _____

    c. _____

    _____

12. Interferences may be of either a positive or negative nature (e.g., erroneously high or low readings). For a given contaminant, would interference problems better be resolved through comparison of:

    a. independent readings from two different infrared analyzers;

    or,

    b. independent readings from a gas chromatograph and an infrared analyzer. Explain.

    _____

    _____

    _____

    _____

    _____

    _____

    _____

    _____

20 Methods of Control

1. Excluding the use of personal protective equipment and administrative controls, list three general means of preventing dust dispersion.

   a. _____

   b. _____

   c. _____

2. Given that small quantities of a contaminant of very low intrinsic toxicity are being released, is it reasonable that general ventilation could be used?

   _____

   _____

   _____

3. What methods of control are actually preferred over local exhaust ventilation?

   _____

   _____

4. If appropriate industrial hygiene concerns were addressed during research and development of processes as well as during the design and commissioning stages, what important steps would be left for industrial hygiene control efforts?

   _____

   _____

5. When they are referenced for industrial ventilation design purposes, it is best to regard the TLVs as:

   a. equal to the design criteria concentrations, (e.g., residual concentrations of airborne materials will approximate the appropriate TLVs).
   b. multiples of the design criteria concentrations.
   c. fractions of the design criteria concentrations.

6. A medical surveillance program in which appropriate biological monitoring is performed can serve as a check of:

    a. engineering and any other controls used.
    b. engineering controls only.
    c. medical controls.
    d. engineering and administrative controls.

7. Which of the following information items are training requirements under the Hazard Communication Standard? (Check all that apply.)

    a. physiological effects of workplace substances
    b. reasons for use of personal protective equipment
    c. plant emergency procedures and policies
    d. hazards of nonroutine tasks

Case Study—Use of Chlorinated Solvents

An industrial hygienist must make recommendations concerning a degreaser that operates using one of several chlorinated solvents. The plant engineer indicates that such devices have been operated in the past using trichloroethylene, perchloroethylene, or methyl chloroform (1,1,1-trichloroethane).

8. Based upon a consideration of the relative toxicities of these compounds, choose one for use in this application based upon health considerations. (These are all toxic materials: central nervous system effects are important at low doses.)

    _____

9. Area sampling (dependent upon the choice of a material indicated in question 1) indicates air concentrations at approximately 330 ppm near this tank. Based upon this information indicate whether or not local exhaust ventilation would be required. Explain.

    _____

    _____

    _____

    _____

    _____

10. Depending upon your choice of solvent in question 1, outline the general features of personal protective equipment necessary for this material. You may assume for this question that some situations will arise in which worker exposures would occur without the use of personal protective equipment. (You may wish to answer this question when you study text Chapter 23, "Respiratory Protective Equipment.")

_____

_____

_____

_____

_____

_____

_____

_____

11. Later, plant engineering wishes to add an arc welding operation approximately 35 feet from the degreaser discussed here. Even though the solvents listed in question 8 are not flammable, locating the degreaser in the vicinity of any "hot" work is strictly prohibited. Why?

_____

_____

_____

_____

21 Industrial Ventilation

1. Flanging a hood opening generally serves to:

   a. increase total airflow.
   b. increase the centerline velocity in the duct.
   c. increase the centerline velocity in front of the hood opening.
   d. decrease capture velocity.

2. The most important factor in the design of a grinding wheel hood is:

   a. taking advantage of particle trajectory.
   b. capitalizing on convective velocity.
   c. taking advantage of slot entry losses.
   d. incorporating a canopy design.

3. The primary purpose of a slot-type hood (incorporating a plenum) is:

   a. providing a canopy configuration.
   b. facilitating total enclosure.
   c. isolating spot currents.
   d. providing air distribution along the slot length.

4. Canopy hoods are most often applied to:

   a. grinding.
   b. hot work.
   c. solvent degreasing.
   d. painting.

5. The control velocity for a canopy hood is maximized when it is positioned:

   a. along a single wall.
   b. in a corner.
   c. in the room's center.
   d. near the ceiling.

6. Given the particle size distribution for a typical fume, the collection efficiency on a number or count basis is (?   ) that on a mass basis. (Choose one.)

   a. greater than
   b. equal to
   c. less than

7. Which one of the following types of air cleaners would not operate with a reasonable efficiency in the "respirable" dust range?

   a. high-efficiency cyclone
   b. electrostatic precipitator
   c. bag house
   d. settling barrel or chamber

8. A cyclone collector can be used upstream of a more efficient air cleaning device primarily as a means of:

   a. improving the collection efficiencies relative to *fines.*
   b. avoiding unnecessary loading of the more efficient (secondary) cleaner.
   c. increasing the overall air velocity.
   d. increasing the overall flow-rate.

9. Check all of the factors listed that are reasonable attributes of a backward-curved fan:

   a. a power demand that reaches a maximum
   b. elimination of the upstream air cleaner
   c. nonoverloading feature
   d. lower coefficient of hood entry

10. In general, when a static pressure check indicates a higher than normal value downstream and a lower than normal value upstream, which one of the following is likely the case?

    a. A blockage has occurred across this span.
    b. A leak has occurred along this span.
    c. The system is operating satisfactorily.
    d. The system should be redesigned.

11. Which of the following would seriously compromise the effectiveness of an open-surface tank ventilation system? (Check all that apply.)

    a. crosscurrents
    b. rapid hoisting of finished materials
    c. inadequate replacement air
    d. nearby window fans being used for relief of heat stressful conditions

12. Given the characteristics of a low-volume, high-velocity exhaust system, which one of the following factors is the most likely problem?

    a. insufficient capture velocity
    b. noise
    c. inadequate duct velocity (permits settling)
    d. inadequate supply (replacement) air

Case Study—Dip-Painting Operation

An industrial hygiene evaluation was conducted at a dip-painting operation to determine if these operations were exposing workers to excessive vapor concentrations. Workers report suffering from headaches, nausea, dry throats, dizziness, and occasional nosebleeds. Interviews with workers revealed the odors were similar to those of airplane glue and/or freshly dry cleaned clothes.

An older local exhaust ventilation system services this operation. It is somewhat similar to that shown in Figure 21-21 in the text, p. 496. Sampling data indicated the following concentrations.

Contaminant	TWA Concentration (Range in mg/m³)
Total alkanes	2.3–267.1
Trichloroethylene	4.0–11.9
Toluene	0.3–0.7
Butyl acetate	1.0–4.2
Xylene	1.8–10.3
Methyl amyl ketone	0.7–2.9
Cellosolve acetate	3.3–23.6

13. Determine whether or not any of these concentrations represent potential overexposures using current ACGIH TLV and BEI criteria. For which of these substances would a TLV for mixtures apply?

14. Predict any change in the effectiveness of this device based upon the following conditions.

    a. increased workload in the plant, necessitating rapid hoisting of painted articles

        _____

        _____

        _____

    b. a shortage of supply or replacement air in the room where this process occurs

        _____

        _____

        _____

    c. a change to the use of a relatively high-order toxicity component in the paint

        _____

        _____

        _____

    d. increased forklift traffic in the immediate vicinity of this tank

        _____

        _____

        _____

15. Suggest a principle of operation for an air cleaner that might be used for this operation. _____

22 General Ventilation

1. What volumetric airflow (cfm) is required to provide six air changes per hour for a room having dimensions of 150 ft × 60 ft × 10 ft?

2. If this room were exhausted through a 3 ft × 3 ft opening, what is the approximate face velocity at the opening?

3. In most cases, replacement air volume should (?   ) the quantity of air removed by (?   ) systems.

   a. equal; exhaust ventilation, process, and combustion
   b. equal; ventilation and process
   c. approximate 90% of; ventilation and combustion
   d. equal; ventilation, process, and combustion

4. Which of the following operations would be best handled using the *air changes per hour* concept?

   a. electroplating
   b. floor sweeping
   c. polyurethane painting
   d. beryllium grinding

5. From the greatest to the least appropriate application, rate the use of the air changes per hour concept.

   a. general offices
   b. photo labs and darkrooms
   c. industrial acid cleaning areas

6. Name at least three types of air supply equipment.

   a. _____

   b. _____

   c. _____

Case Study—Replacement Air in an Industrial Park

An industrial park recently added a refuse-derived fuel (RDF) plant to its tenants. The RDF plant will be able to provide heating gas at a very reasonable price for all park occupants. Previously, other tenants have relied on oil-fired heaters.

7. What are some of the major issues related to this changeover to a new fuel supply? (This question is asked only in terms of replacement air considerations.)

   _____

   _____

   _____

   _____

   _____

   _____

   _____

   _____

   _____

   _____

   _____

23 Respiratory Protective Equipment

1. Which of the following agencies never had regulatory responsibility for official respirator testing protocols?

   a. U.S. Bureau of Mines
   b. Mine Safety and Health Administration
   c. Occupational Safety and Health Administration
   d. National Institute for Occupational Safety and Health

2. Name five elements of a minimum respiratory protection program.

   a. _____

   b. _____

   c. _____

   d. _____

   e. _____

3. Certain organometallic compounds, such as zinc resinate, are added to respirator dust filters during the manufacturing process. What does the addition of such a static-charged metallic compound accomplish when metal fumes pass across the filter?

   _____

   _____

4. Why is a high-temperature alarm necessary on an oil-lubricated compressor?

   _____

   _____

   _____

   _____

5. The medical screening portion of a respiratory protection program is aimed primarily at:

   a. identifying individuals potentially sensitive to the toxic exposures involved.
   b. identifying individuals likely to be less tolerant of the additional stress associated with respirator use.
   c. determining the pulmonary capacity of respirator users by pulmonary function testing.
   d. evaluation of the adequacy of protection provided by the respiratory protective device.

(The following three questions simulate practical assignments regarding selection and use of respiratory protection. In each case, form an opinion concerning the adequacy of the device in use. Use ACGIH TLVs for reference values. See "Appendix B-1").

6. A laborer complains of dust to the superintendent, who supplies him with a single-use, disposable respirator, which does not bear any approval. The shop steward tells the superintendent that this is a violation of the OSHA regulations. Respirable (quartz) silica concentrations approximate 0.02 mg/m³; while nuisance dust concentrations are at 4 mg/m³. Is this a violation?

_____

_____

7. A silk screen craftswoman performs her craft without any local exhaust ventilation or respiratory protection. Sampling and analysis reveal she is being exposed to approximately 5 ppm of isophorone. What type of respirator would be required?

_____

_____

8. A maintenance worker assigned to clean a large degreaser apparatus must occasionally go into the pit area. He complains that the odor of 1,1,1-trichloroethane is strong when the vapors come through his respirator cartridges. Should he be assigned another type of respirator with a higher protection factor, or can cartridge changes solve this problem?

_____

_____

9. Of the following forms of personal protective equipment, which class of devices would have the poorest worker acceptance?

    a. air-purifying respirators
    b. hard hats
    c. safety glasses
    d. insert-type hearing protection

10. Which one of the following statements best describes an employer's obligation to use administrative or engineering controls instead of respirators?

    a. Personal protective equipment is a substitute.
    b. Personal protective equipment is an alternative, when engineering controls are not feasible.
    c. Personal protective equipment should not be used.
    d. Respirators are an easy fix for most problems.

11. Which of the following respirator types suffer(s) most from inward leakage?

    a. air-supplied respirators
    b. air-purifying respirators
    c. self-contained breathing apparatus
    d. air-line devices in pressure-demand mode

12. Why can't a half-mask respirator be used for materials such as gaseous ammonia or sulfuric acid? _____

    _____

    _____

    _____

13. Given that gas masks are not approved for concentrations exceeding 2% (v/v), is it realistic to advise their use for hazardous entry? Can such devices be worthwhile for emergency exit?

    _____

    _____

    _____

    _____

14. Do gas masks and cartridge-type half-mask respirators appropriate for a given class of contaminants differ in terms of the chemical composition of the cartridges? Do they differ in the amount of adsorbent material? Can any of these devices ever be used in oxygen-deficient atmospheres?

_____

_____

_____

Case Study—Crop Dust Flagman Exposure

A flagman for an aerial crop duster is spraying Mevinphos dust (2-carboxy-1-methylvinyl dimethyl phosphate). The pertinent OSHA PEL is 0.1 mg/m³, and the pesticide is immediately dangerous to life and health at a concentration of 40 mg/m³.

Vapor pressure = 0.003 mmHg (at 20 C)
Flash point = 175 C.
Mol. wt. = 224.1

After using a single-use disposable dust respirator, the flagman experiences tightness of the chest, wheezing, tearing, and a running nose. Sweating and itching along the forearms and hands are pronounced. He also exhibits constricted pupils.

15. This information emphatically suggests that the flagman is being overexposed. Assuming that no sampling data are available, use the vapor pressure to calculate a possible worst case vapor concentration. This may represent any number of actual work situations. Reference text Chapter 17 for help in performing the calculation.

16. Discuss the appropriate means of both respiratory and whole-body protection necessary for this individual, if sampling indicates his exposures approximate 10 mg/m³.

Case Study—Chemical Processing Plant

A survey was conducted regarding potential occupational exposures to bladder carcinogens in a chemical processing plant. Agents of particular interest included 3,3'-dichlorobenzidine (DCB), o-dianisidine (ODA), and o-toluidine. Samples were taken:

a) inside a reusable air hood of a two-piece full-body suit (note: the hood and suit are joined together to make a seal);
b) in nonregulated areas adjacent to a charge room;
c) along a forearm gauze patch underneath workers' coveralls;
d) from surfaces routinely contacted by workers.

Given the following three hypothetical situations, outline any problems involved or state that the protection appears adequate from the information provided. Remember that deficiencies can occur with personal protective equipment besides respiratory protective devices.

17. Samples inside the airspace of the reusable air hood are significant, while forearm gauze patch samples do not indicate levels above those of background.

18. Samples inside the airspace of the hood are insignificant while both forearm gauze patch samples and wipe samples taken from room surfaces are elevated.

19. Levels of DCB in the urine of exposed workers are significant; levels inside the airspace of the hood are significant; levels along the gauze patch samples also are elevated.

24 The Industrial Hygienist

The case studies and other questions relative to Part Six, Indutrial Hygiene Programs, are general in nature and do not require specific answers. Instead, these exercises are catalysts for discussion. Where appropriate, specific answers have been provided but otherwise it is best to consult the reading material for guidance. This section better reflects the "art" rather than the science of industrial hygiene.

Your company, a consulting firm previously engaged in air-pollution engineering studies, is broadening its services to include the industrial hygiene area. Anticipated industrial hygiene services include testing relative to asbestos abatement, indoor air pollution, hazardous waste cleanup, and primary metal production.

Shifting business patterns allow a junior air pollution (B.S., Mechanical Engineering) engineer to be assigned to this new endeavor. Management also contemplates hiring four more individuals at various grade levels to begin this program.

Assume the following: (1) Your duties are strictly managerial and you are not permanently assigned to this group; (2) a trade-off exists between financial commitment to personnel and capital outlays for equipment; (3) within 30 months, revenues from the new industrial hygiene services must match and then exceed basic personnel and equipment costs. Facility costs are not an issue.

Given this scenario prepare *background* materials relative to these areas.

1. For the four new positions, list basic position requirements (e.g., prerequisite skills, education, and responsibilities for the positions). A range of alternatives are appropriate but each must be justified.

_____

_____

_____

_____

_____

_____

_____

_____

_____

_____

_____

_____

_____

_____

_____

_____

_____

2. For the mechanical engineer outline a three-year developmental plan that will allow him to reasonably sit for the industrial-hygienist-in-training examination at the end of this period. For purposes of this exercise, assume that management is willing to allow this individual 25 in-service days per year for training activity.

_____

_____

_____

_____

_____

_____

_____

_____

_____

_____

_____

3. Using information provided in chapters 19 and 20, list equipment that is essential for this venture. A short written justification should accompany each item.

_____

_____

_____

_____

_____

_____

_____

_____

_____

_____

(Instructors may wish to add appropriate dollar ceilings to this exercise and to make any number of other provisions.)

(Questions 4 and 5 do not relate to the passage.)

4. What are the ultimate goals of the basic sciences (e.g., biology, chemistry, physics, and toxicology) and specialized training for industrial hygienists?

_____

_____

_____

_____

_____

5. According to information presented in the text and your opinions, which of the following is more feasible?

   a. To provide industrial hygiene training for a personnel administrator who is knowledgeable regarding company policies (who will then perform industrial hygiene functions);
   b. To provide training for an entry-level industrial hygienist (B.S., Industrial Hygiene) relative to company policies.

_____

_____

_____

_____

(Please note: There are no specific answers to questions on Chapter 24.)

25 The Safety Professional

1. While the goals of the safety profession may be largely directed at accident prevention, substantial overlap exists with industrial hygiene program goals. Industrial hygiene work is typically divided into phases of anticipation, recognition, evaluation, and control.

   Use these four categories to group the accident prevention activities listed on p. 590 in the text. In your opinion, which of these safety-associated activities are most strongly correlated with industrial hygiene program activities?

   _____

   _____

2. Does the purpose of a safety inspection change depending upon whether it is conducted by plant or third-party personnel? What advantages and disadvantages exist for a third-party inspector?

   _____

   _____

3. Define the following:

   a. job safety analysis: _____

   _____

   b. loss control: _____

   _____

   c. systems safety: _____

   _____

   d. fault-tree analysis: _____

   _____

(Please note: There are no specific answers to questions on Chapter 25.)

26 The Occupational Physician

As corporate director of environmental health and safety, you are responsible for supervising services offered in a broad range of areas, including occupational medicine. While you were gone, your secretary scheduled an appointment with the corporate physician, who reports to you, and who wishes to discuss a number of issues. Before the meeting you need to acquaint yourself with the following agenda items. What notes would you make before the meeting?

1. At a rural Wisconsin plant, the physician, who has provided medical emergency support, is retiring from private practice and a replacement must be found.

_____

_____

_____

_____

_____

_____

_____

_____

_____

_____

_____

_____

2. At one of your plants in Mexico, a considerable number of workers are acquiring tuberculosis (TB). The local health authorities have not been aggressive and issues exist whether the plant should provide TB testing for workers—Their families—Contacts and other members of the community.

     _____

     _____

     _____

     _____

     _____

     _____

     _____

3. Community right-to-know legislation requires revision of a very brief, and probably inadequate policy statement concerning emergency care for employees of an oil refinery.

     _____

     _____

     _____

     _____

     _____

4. Advances in biological monitoring suggest that employee health records should in some cases have new entries regarding both air exposure and biological monitoring data.

     _____

     _____

     _____

     _____

     _____

5. Workers at a plant in Texas are asking to review their medical records. The plant has been repeatedly cited by OSHA for violations of the lead standard.

_____

_____

_____

_____

_____

Case Study—Foundry Workers

Workers in a nonferrous foundry requested a combined medical and industrial hygiene evaluation on account of alleged adverse health effects; dermatitis and upper respiratory irritation, presumably from exposure to fluorides and pneumoconiosis from exposure to carbides and diborides.

Questionnaires, pulmonary function tests, and chest x-ray films were used to determine possible respiratory effects from carbides and diborides. Questionnaires and chest x-ray films of workers potentially exposed to these two classes of materials compared favorably to those of a control group, while results of common pulmonary function tests appeared abnormal.

The potential health effects of fluorides were also evaluated via questionnaire and measurement of urinary fluoride levels. Urine fluoride levels were consistent with established values for working populations but were higher in the exposed workers than in a nonexposed group. Physicians also noted nasal irritation and slowly healing cuts in exposed workers.

6. Using information provided in this chapter and also in Chapter 2, "The Lungs," describe several types of pulmonary function tests that would be potentially useful in this example.

_____

_____

_____

7. In the assessment of total exposure to fluorides, what advantage does biological monitoring (urinary fluoride levels) have over more traditional methods such as air-sampling?

_____

_____

_____

_____

_____

_____

_____

8. What problems would exist in interpreting urinary F-concentrations, if there was no BEI for fluoride available?

_____

_____

_____

_____

_____

_____

_____

27 The Occupational Health Nurse

At a southeastern textile plant, your section includes one safety manager, an industrial hygiene technician, and three industrial health nurses. Two nurses are registered, possessing baccalaureate degrees, while one nurse is a licensed practical nurse. In addition to other nursing-related duties (e.g., provision of first aid services, health education, employee assistance programs), the nurses have been tasked with rapidly conducted audiometric screening and pulmonary function testing for appropriate members of the 900-employee work force. Employees staff two shifts on a seven-day-a-week basis.

It may help to know that the consulting plant physician practices internal medicine in a nearby community. Patients are referred to his clinic; he has visited the plant three times in the last ten years.

The safety professional reports to the plant manager, while the technician reports to the corporate industrial hygienist, who is located at corporate headquarters in another part of the county. The safety manager provides most of the direction for the nurses except clinical direction, which may be provided by the physician. No nursing-specific lines of authority exist.

Alternatives in terms of completing the audiometric and pulmonary function testing programs within a two-month period include the following:

a. Complete in-plant provision of services by the nursing staff. Nursing staff members currently possess knowledge and certifications to do these tests;

b. Contract services by a vendor. The vendor is a reputable provider of occupational health-related services;

c. In-plant provision of testing by the nursing staff with the assistance of the industrial hygiene technician and two summer employees (college nursing students). The nursing service must provide training for all of the technician employees.

1. What are the major issues when selecting alternatives with respect to:

   a. quality assurance _____

   _____

   _____

   b. costs _____

   _____

   c. ability to continue testing at frequent intervals _____

   _____

   d. proper time management for all staff involved _____

   _____

2. Does the organizational structure described seem conducive to effective communications among health and safety professionals? How could you improve upon the organizational structure? Provide a flow diagram.

   _____

   _____

   _____

3. Do the proposed assignment of duties appear consistent with the goals and objectives of occupational health nursing?

   _____

   _____

   _____

   _____

(Please note: There are no specific answers to questions on Chapter 27.)

28 The Industrial Hygiene Program

Retirement of the company's senior industrial hygienist has resulted in your promotion to that rank. Although you are recognized by your co-workers as occupying the senior industrial hygiene position, top management has never designated this. Such a designation exists for the safety professionals, however.

The issue is not new, but due to the retirement and other company changes, your supervisor, has agreed to arrange a 45-minute meeting with him, his supervisor, the corporate medical director, and two key plant managers to discuss the issue. Restructuring would allow both safety and industrial hygiene directors to report to the corporate medical director on a parallel basis. The historical reluctance to place the industrial hygiene function at a higher level has stemmed in large part from a lack of evaluation of industrial hygiene program effectiveness.

1. Using the text Chapter 28, "The Industrial Hygiene Program," write a general outline that explains key points regarding improved program evaluation.

_____

_____

_____

_____

_____

_____

_____

_____

2. Substantial clerical and database resources for program evaluation reside in the safety directorship but certain advantages would also attend an independent industrial hygiene function. Justify alternative plans for (a) retaining the present lines of authority; or (b) granting directorship status to industrial hygiene.

_____

_____

_____

_____

_____

_____

_____

(This question does not relate to the passage.)

3. To establish an industrial hygiene program, Chapter 28 lists these four components: (1) recognition, (2) evaluation and (3) control of health hazards, and (4) employee education and training. What are the advantages of using this four-part approach?

_____

_____

_____

_____

_____

_____

(Please note: There are no specific answers to questions on Chapter 28.)

29 Computerizing an Industrial Hygiene Program

1. What are the main considerations of a requirements study, to be conducted before the purchase of a computer system for an industrial hygiene program?

_____

_____

_____

_____

_____

_____

_____

_____

_____

_____

_____

_____

_____

_____

_____

_____

_____

2. One of the key issues facing information systems is the ability to record and integrate data of a scheduled or an unscheduled nature. Provide at least two examples of each condition as they refer to occupational health records.

_____

_____

_____

_____

_____

_____

_____

_____

_____

3. Define the following:

a. LAN: _____

_____

_____

b. Needs analysis: _____

_____

_____

c. System manager: _____

_____

_____

(Please note: There are no specific answers to questions on Chapter 29.)

30 Governmental Regulations

Your firm, a large multistate battery manufacturer, was just cited by OSHA for unhealthful lead exposures and certain other deficiencies. You have recently been reassigned to this facility after your predecessor retired from the company. Since your arrival you have begun a thorough and ambitious program to try to improve working conditions, but it is obvious "that Rome won't be built in a day!"

Your firm was cited for failure to maintain (and control) airborne lead levels below the PEL, failure to perform medical surveillance for certain employees who are being exposed at or above an action level, and deficiencies in its Hazard Communication Program.

(This series of questions will test your knowledge of legal requirements relative to the Occupational Safety and Health Act and pursuant regulations.)

1. In terms of developing an improved occupational health program for this facility, why is the term "action level" significant? How can action level criteria effectively be used in your discussions with plant management?

_____

_____

_____

_____

_____

_____

2. If the citation was mailed on the sixth of the month, received, and stamped in your plant on the tenth of the month (a Monday), how long do you have to respond, if you plan to contest the citation?

_____

_____

3. Name at least five features of employee training specified by the Hazard Communication Standard that must be addressed.

    a. _____

    _____

    b. _____

    _____

    c. _____

    _____

    d. _____

    _____

    e. _____

    _____

4. Does effective medical surveillance for lead mean that all plant employees must be placed in a blood-lead screening program? (You may wish to consult text Chapter 26, "The Occupational Physician," regarding this question.)

    _____

    _____

    _____

    _____

    _____

    _____

(The following questions do not relate to the preceding information.)

5. By legislative fiat, the primary responsibility for training of employers and employees rests with:

    a. NIOSH.
    b. OSHA.
    c. Occupational Safety and Health Review Commission (OSHRC).
    d. ACGIH.

6. Which one of the following agencies has primary responsibility for conducting educational programs to ensure an adequate supply of qualified safety and health professionals?

   a. NIOSH
   b. OSHA
   c. OSHRC
   d. ACGIH

7. Employers cited by OSHA may seek recourse by:

   a. appealing to OSHA and the OSHRC.
   b. requiring OSHA to repeat the inspection.
   c. appealing through the courts.
   d. more than one of the choices listed above

8. Which of the following is the agency with primary responsibility for standards promulgation?

   a. NIOSH
   b. OSHA
   c. Congress
   d. *Federal Register* Publisher

9. The relationship between the Department of Labor and the Occupational Safety and Health Review Commission can best be described as:

   a. The two are completely independent.
   b. The OSHRC reports to the Assistant Secretary of Labor.
   c. The OSHRC reports to the Secretary of Labor.
   d. A choice not listed here

10. Name two obligations of employers and one general obligation of employees pursuant to the "General Duty" clause.

    _____

    _____

    _____

    _____

31 OSHA: The Federal Regulatory Program— A History

Few observers of the signing of the Occupational Safety and Health Act could have anticipated the direction that agencies and major regulations authorized by the Act would take in the next two decades. This history has important implications for the future of occupational health and safety and for other regulations designed to protect the public. The following questions should allow the reader to assess future developments.

1. Pick any two Assistant Secretaries of Labor for OSHA, and question whether their policies significantly affected the OSHA, or whether events that were largely outside of their control served to shape the major course of action during their tenure.

_____

_____

_____

_____

_____

_____

_____

_____

2. One of the major criticisms of OSHA has been its inability to protect employees in many small businesses. Is there anything in the agency's history or in its current development that would suggest improved strategies?

_____

_____

_____

_____

_____

_____

_____

3. What has been done during nearly 20 years of OSHA's existence to broaden worker participation and to ensure that every worker knows about pertinent occupational health and safety information?

_____

_____

_____

_____

_____

_____

4. How is an Executive Order used to extend safety and health protection to federal employees?

_____

_____

_____

_____

_____

5. Compare and contrast generic versus substance-by-substance rulemaking.

_____

_____

_____

_____

_____

_____

_____

_____

(Please note: There are no specific answers to questions on Chapter 31.)

Chapter 1—Answers

1. a, b, d, and e (These individuals have professional training that qualifies them.)

2. c

3. c (We want to prevent exposures.)

4. d

5. a

6. d (Answer derived from TLV information, too.)

7. a–d (See p. 10 in text.)

8. c

9. a

10. c

11. a

12. b and c (Answer "a" has an entirely different meaning.)

13. All are chemical agents except for microbial contamination (biological agents) and heat stress (physical agent).

14. See pp. 6–21 in text.

15. Published recommendations include the TLVs; NIOSH criteria might also be used; OSHA PELs constitute regulated concentrations.

16. Medical surveillance data; engineering data; interviews with workers.

Chapter 2—Answers

1. d

2. Several approaches can be used. One simple approach is to use the 24-hour value given in text p. 40.

$$\frac{(12{,}491 \text{ L}) \ (8 \text{ hr})}{(24 \text{ hr})} = 4122 \text{ L (approximately } 4.1 \text{ m}^3)$$

3. a

4. c

5. b

6. c

7. d

8. d

9. b

10. b (text Figure 2-13, p. 40).

11. 5%

12. a

13. See explanation on p. 36 in text.

14. Material from text Chapter 1 about sampling can be mentioned, and pulmonary function testing can be outlined.

15. Mucus, cilia, turbinates, and progressive narrowing of the branches of the respiratory system.

16. See flour mill dust (text Figure 7-2, p. 126–10 μm).

17. The answer to the first part of question 17 can be found in Table 7-E, pp. 142-143 in text. Presence of any inert material, oxygen concentration.

Chapter 3—Answers

1. a. Lightly-pigmented skin is more susceptible to UV radiation.
   b. UV radiation contributes to the photosensitization phenomenon.
   c. Sweat gland disorders make heat difficult to dissipate.

2. e

3. d

4. b

5. a

6. d

7. b

8. The limiting nature of the stratum corneum should be mentioned; effect of broken of disease skin should be noted; see p. 51 in text for other points.

9. Mechanical trauma

10. Biological and chemical agents. The biological agents are fairly obvious, and include enteric viruses, infectious bacteria, and fungi. These workers are also at risk from numerous chemical agents that are part of such effluents; pesticides immediately come to mind.

11. Chromium and arsenic compounds used in leather tanning

12. Varying thicknesses, the palm is callused, whereas the back of the hand is not.

Chapter 4—Answers

1. d

2. e

3. a

4. a–d

5. a

6. a–d

7. a–d

8. b

9. c

10. Temporary threshold shift

11. Noise-induced hearing loss; some workers will still lose hearing even when exposed to as little as 80 dBA. (It may be necessary to consult text Chapter 9, "Industrial Noise," to get a better idea.)

12. 85 dBA

Chapter 5—Answers

1. Cornea, lens

2. d

3. Bending of light rays

4. Choroid

5. a. Lack of stereopsis—others may be important but depth perception is foremost.

6. d

7. Flexible-fitting goggles or chipping goggles; current ANSI Standard (Z87 series).

8. No answer required.

9. Calcium hydroxide

10. Foreign object risk

Chapter 6—Answers

1. e

2. b

3. c. Hydrogen sulfide (H_2S) is a gas at room temperature and pressure.

4. a

5. b. Note the difference between questions 1 and 5. Question 5 asks which is most susceptible, it does not infer that in every case any organ system will be affected.

6. b. Calculations generally indicate that the TLVs are at least an order of magnitude less than the lower explosive limit. Consider a flammable solvent having a lower explosive limit of 1% (which translates to 10,000 ppm v/v). Such a solvent would likely have a TLV less than 1,000 ppm.

7. c. The mist could give rise to an airborne glycol compound; answer d. might merit some attention as well, if hoist speeds were not limited.

8. a

9. a

10. c

11. d

12. c

13. a, b, d. Air is being monitored so a check against liquid concentrations would not be useful.

14. a–d

15. b

16. The liquid with the lowest flash point

17. The means must include the bonding and grounding provisions discussed in the text; examples could range from portable safety cans and suitable dispensing stations to drums equipped with rotary pumps.

18. Toluene and benzene are both aromatic hydrocarbons. According to Table 6-C, p. 108 in text, PVA has the best characteristics; all other entries in the table are unsuitable.

19. Vapor hazard ratios for toluene (368) and ethylbenzene (126); ethylbenzene would be preferred on this basis. Results can be checked against Table 6-B, p. 104 in text.

20. Obviously, workers who are ill and away from the workplace cannot be monitored in a typical manner; some type of interviews and/or review of their medical records must be conducted. Historical data on related cases might be consulted as well.

Chapter 7—Answers

1. c

2. b, c, and d

3. a

4. d

5. d

6. a, b, and c

7. a. Uncomplicated asbestosis is a nonmalignant condition; the patient may develop a malignancy and is at increased risk compared with normal populations.

8. Siderosis (iron oxide and silica)
   Berylliosis (beryllium)
   Coal workers' pneumoconiosis (coal dust, silica)
   Kaolinosis (china clay)
   Bauxite pneumoconiosis or Shaver's disease (aluminum oxide, silica)
   Other answers possible

9. d. chrysotile (the only serpentine form)

10. (1988 ACGIH TLV/BEI Book) Cristobalite and tridymite (0.05 mg/m³); other forms (0.10 mg/m³)

11. quartz

12. d

13. Text Figure 7-4, p. 137, shows the cyclone assembly being used with a personal sampling pump

14. Text Figure 7-4, p. 137, indicates (1) properly positioning the sampler on the worker; (2) the use of a rotameter (secondary standard); (3) recording start and stop times; (4) checking the flow; and (5) other steps that may be noted in this answer, as well.

Chapter 8—Answers

1. a

2. a

3. a

4. d

5. a

6. a

7. c

8. This is very probably irritant contact dermatitis, given the presentation shortly after exposure; also, the fact that such a large percentage of workers are affected confirms this finding.

9. This finding is consistent with patterns of clothing use during warm months; as workers shed clothing, dermatitis occurs on exposed skin surfaces such as the forearms.

10. Proper planning, process control and selection of materials would head the list of control steps. Measures such as good housekeeping and the use of personal protective equipment, would not rank as high.

11. Irritant and allergic contact dermatitis should be investigated.

12. The barrier cream might be considered as a supplement to the steps outlined in the answer to question 10, depending upon the quality of documentation regarding its tested effectiveness. However, it still ranks well down the list of alternatives within the basic category of personal protective equipment.

13. No specific answer is necessary.

Chapter 9—Answers

1. b

2. b

3. b

4. a

5. a

6. a

7. a. See text Figure 9-8, p. 170.

8. A-weighting network.

9. b

10. c

11. a

12. b

13. d

14. a

15. a

16. $(1/6.0) + (3/16.4) + (4/9.2) = 0.78$

Period	Period (hr)	SPL (dBA)	Permitted Duration
1200–1300	1	92	6.0
1301–1600	3	84	16.4
1601–2000	4	89	9.2

(Table G-16A should be consulted, p. 195.)

17. Decibel addition must reflect the logarithmic nature of the measure. When levels differ by 5, 1.2 dB is added to the higher level and a result computed $(98.0 + 1.2 = 99.2)$; when results differ by 10, 0.4 dB is added to the higher $(99.2 + 0.4 = 100.2)$.

18. Given an overall reading of 100.2 dBA with all three devices operating and a reading of 98 dBA with only the No. 5 Pug Mill operating, the reading could never be less than 98 dBA if the other two devices were eliminated, much less just controlled.

Chapter 10—Answers

1. a and d

2. a and c

3. c and e

4. a

5. a

6. b

7. b

8. b

9. a

10. b

11. X-ray

12. The half-life for Cs^{137} is 30 years; the isotope is long-lived, hence, contamination can be significant for years.

13. The exposure rate is calculated using the inverse square law, e.g.:

    exposure rate   = (25 rem/hr)(25m/40m)^2
    inspector B

                      = (25 rem/hr)(0.39) = 9.75 rem/hr

14. milliroentgens/hr, x-, or gamma-rays

15. Contaminated biological systems pose a risk to others; irradiated systems do not directly pose a risk to others.

Chapter 11—Answers

1. Burns, cataracts, possible reproductive damage

2. c

3. The weighting formulae shown below should be used as indicated in Table 16 (Relative Spectral Effectiveness by Wavelength) of the current ACGIH TLV/BEI Booklet.

$$E_{eff} = \Sigma \, E_\lambda \, S_\lambda \, \Delta_\lambda$$

4. Quantity as indicated by illuminance, quality as indicated by an absence of glare, proper diffusion, direction, color, brightness, and other factors

5. (1) cosmic and gamma rays, (2) x-rays, (3) ultraviolet, (4) visible, (5) infrared, (6) microwaves, (7) radio, (8) extremely low frequency

6. Many choices are possible; material is distributed throughout the text chapter.

7. d

8. a

9. c

10. b

11. c

12. 1 mW/cm²

13. The TLV would be different, about 3 mW/cm².

14. Infrared-retina

15. Yes

16. Duration of exposures, size of laser beam (narrow band or extended source), information describing average power output, peak power, pulse duration, and repetition frequency, laser-source radiance, or integrated radiance

17. No, protective eyewear is reasonably specific to particular wavelengths. Since the three devices operate at three different wavelengths, only one device should be operated at a time.

Chapter 12—Answers

1. M is always positive; E is always negative.

2. 38 C

3. d

4. c

5. b

6. a

7. d

8. c

9. In general, for new workers the heat stressful exposure is limited to 20% on the first day, with 20% increments on successive days; seasoned workers may tolerate an accelerated regimen.

10. d

11. a

12. Heatstroke, heat syncope, heat cramps, heat rash, others as listed in text Table 12-B, p. 267

13. This point corresponds well to the 50% work/50% rest line.

14. Not good for pointing out causes of heat stress or controls that may be necessary

15. b. Notice how severe these conditions are.

16. (Per text Figure 12-10, p. 275). HSI Calculations: Enter A using intersection of t_g and air speed; drop vertical line into B, intersecting with metabolism; draw horizontal line into HSI Block and then intersect t_{db} and t_{wb} in lower left psychrometric chart; draw horizontal line to Y, intersect with air speed; and draw vertical line to intersect horizontal line in HSI block:

$$HSI = 140$$

WBGT Calculations:

$$WBGT = 0.7t_{wb} + 0.3\ t_g = (0.7)(89) + (0.3)(100) = 92$$

17. ET does not take metabolic heat production into account; therefore, it is primarily applicable to sedentary workers.

Chapter 13—Answers

1. a

2. Standing (330 N); seated with arms hanging (500 N)

3. d

4. c. Odd as it might seem, a *single fixed* seat height at the the 50th percentile can accommodate a very minor portion of the population; to accommodate a larger fraction, a range of seat heights must be specified.

5. c

6. a

7. b

8. b

9. a

10. The statement is false, human comfort considerations must be incorporated into the original design for there to be a reasonable chance of success.

11. b. A broader or more limited range might be used, but generally this range proves satisfactory.

12. Combinations could include glare from the VDT screen (stress), which leads to eye fatigue (strain); job-related (stress), which leads to chest pain or angina (strain); uncomfortable workstations (stress), which leads to physical malady (strain). The key point is that the stress is external while the strain is the human manifestation.

13. Generally, it appears the strain associated with VDT use does not occur as much from unsafe levels of electromagnetic energy (x-rays, UV, IR, others) as much as it arises on account of secondary factors (e.g., job-related stress, uncomfortable workstations, ergonomic problems, indoor air pollution associated with office environments, or high levels of noise).

14. c

15. The answer to this question requires a general reading of the chapter. Mentioned in various sections are interview techniques, biomechanical examinations, medical evaluation, and other techniques. (Many answers are acceptable.)

16. Solutions could range from redesign of the workstation to administrative controls of various types (e.g,, restricting manual materials handling to only certain operations).

Chapter 14—Answers

1. All of these choices could be important; the chapter contains ample material to justify each.

2. Per text Table 14-C, p. 340:

a. Brucellosis:	Cattle workers
b. ORF:	Slaughterhouse workers
c. Histoplasmosis:	Poultry handlers
d. Leptospirosis:	Sewer workers, coal miners
e. Hepatitis B:	Persons providing medical or dental care

3. Many answers are possible. Choices that quickly come to mind are bloodborne diseases, such as hepatitis for the biomedical technicians; Pontiac fever and legionnaires' disease for industrial hygienists; any number of bioaerosols for the dental personnel.

4. b

5. b

6. d

7. Zoonoses represent diseases transmissible from vertebrate animals to humans and other animals. Given the animal research being conducted in the buildings, the animals could represent a reservoir for disease.

8. Inhalation of fungal or actinomycete spores can lead to this diagnosis.

9. Poor maintenance of the system could mean that antimicrobial agents were not used and that slime was allowed to accumulate; the flood would indicate some high humidity conditions and a means by which liquid wastes could be transported.

10. Class I, since no concern exists with respect to contaminating the blood or urine samples with organisms from room air.

11. The biological safety cabinets do not offer suitable protection against chemical contaminants while the chemical hoods are inadequate for biological agents.

Chapter 15—Answers

1. Both morphological (liver and kidney) and functional (nasal and bronchial irritation) criteria are noted.

2. Irritation

3. The irritation response is definitely applicable to humans as well as animals. The liver and kidney changes are indeterminant.

4. Probably so, even though irritation may represent the acute response, liver and kidney changes may occur on a chronic basis.

5. The presence of a dose-response relationship simply means that increases in dose will influence biological response in some systematic manner. A variety of curves (see text Figure 15-3, p. 363) are possible; empirical studies are typically the only means of verifying the relationship.

6. Chronic—The length of the study is consistent with a worker's lifetime risk.

7. The data are certainly useful although incomplete. More information is necessary concerning the potential liver and kidney damage. Sex differences should also be investigated in greater detail.

8. a, b, and d

9. a

10. a

11. d

Chapter 16—Answers

1. a and b

2. a

3. d

4. c

5. a

6. b

7. b

8. The continuous monitoring might avert the coordination problem encountered when sampling during the production runs. Estimation of personal exposures might be deficient; area samples typically do not provide a good estimate of personal exposures.

9. The biological monitoring could be tied to the most common exposures (by any route). In particular, exposures to solvents might be estimated using biological monitoring techniques. Because other means of sampling have shortfalls, biological monitoring could serve as a useful adjunct.

Chapter 17—Answers

1.
$$\text{Concentration (ppm)} = \frac{400 \text{ mmHg } (10^6)}{(760 \text{ mmHg})} = 5.26 \times 10^5 \text{ ppm}$$
$$\text{Concentration (mg/m}^3) = 5.26 \times 10^5 \text{ ppm} \frac{(100)}{(24.5)}$$
$$= 21.5 \times 10^6 \text{ mg/m}^3$$

2.
$$\frac{25}{10^6} = \frac{\text{unknown CCl}_4 \text{ volume (g)}}{5 \text{ L (air)}}$$
$$\text{CCl}_4 \text{ (g)} = 1.25 \times 10^{-4} \text{ L}$$

At 25 C and 1 atm, 1 mole of ideal gas occupies 24.5 L
Therefore:
$$\frac{1 \text{ mole}}{24.5\text{L}} = \frac{\text{unknown moles}}{1.25 \times 10^{-4}\text{L}}$$
$$\text{Unknown moles CCl}_4 = 5.1 \times 10^{-6}$$
$$\text{Volume liquid CCl}_4 = (5.1 \times 10^{-6} \text{ moles}) \left(\frac{154 \text{ g}}{\text{mole}}\right) \left(\frac{1 \text{ mL}}{1.5 \text{ g}}\right)$$
$$= 5.24 \times 10^{-4} \text{ mL (liq.)}$$
$$= 0.52 \text{ } \mu\text{L}$$

3.
$$\text{TWA} = \frac{(75)(320\text{ppm}) + (100)(250\text{ppm}) + (230)(350\text{ppm})}{(405)}$$
$$\text{TWA} = 319 \text{ ppm}$$

4. a. area; b. personal

5. The use of (b) compressed air or (d) steam represents a significant hazard because these techniques tend to disperse aerosols of contaminants; (c) vacuuming and (a) wet mopping come with better recommendations because they are not as likely to do this. In the case of vacuuming, however, questions arise regarding the filters that are used in the device; if the filter is inadequate the vacuum itself may redisperse particles.

6. Consult text Table 17-B, p. 399.

7. a. exceed the action level but do not exceed the permissible exposure limit. Control efforts are necessary; resurvey may be necessary as well.

   b. exceed the permissible exposure limit. Control efforts are necessary, halting the operation should be seriously considered if controls cannot be used. Follow-up surveys are necessary.

   c. approximate 1/100 of the action level. No action is probably necessary unless special conditions exist (carcinogen, teratogen, etc.). Keep in mind, however, that all unnecessary exposures should be avoided.

8.
$$\text{Concentration} = \frac{\text{Amount}}{\text{Volume}} = \frac{100\ \mu g}{1 m^3} = \frac{25\ \mu g}{0.25 m^3} = \frac{25\ \mu g}{250 L}$$
$$\text{Sampling Rate} = \frac{2.5\ L}{\min} = \frac{250\ L}{100\ \min}$$

9 a. Understand the limits of sampling and analysis methods; (b) properly calibrate and maintain equipment; (c) develop or use both internal and external quality assurance programs.

10. a

Chapter 18—Answers

1.

Midget impinger

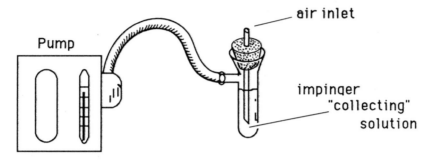

Mixed fiber filter

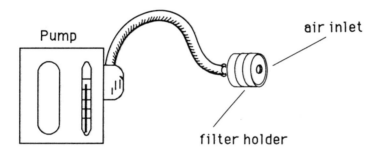

Adsorbent tube

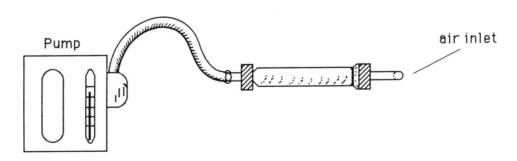

2. d

3. c

4. a

5. b

6. b

7. b

8.

Trial	Time (sec)	Volume (mL)	Flow-rate (mL/sec)
A	22	720	32.7
B	23	750	32.6
C	21	680	32.3
		Mean =	32.5 mL/sec or 1,950 mL/min

9. $\text{Concentration} = \dfrac{\text{Amount}}{\text{Volume}}$

$= \dfrac{(0.1994 - 0.1280\text{g})}{(32.5\text{mL/sec})(400\text{min})(60\text{sec/min})}$

$= (9.15 \times 10^{-8} \text{ g/mL})$

$= (9.15 \times 10^{-8} \text{ g/mL}) (10^3 \text{ mg/g})(10^6 \text{ mL/m}^3)$

$= 91.5 \text{ mg/m}^3$

10. a. The objective of the sampling—Since concern about cardiovascular disease exists, the exposure monitoring must address this question. Personal sampling is indicated.

   b. Interference questions—Given that several metals are present, can the sampling/analysis scheme distinguish among them?

   c. Questions regarding sensitivity—Will the method be able to detect low concentrations of these substances. These and other concerns are generally addressed on p. 418 in text.

Chapter 19—Answers

1. a. ultraviolet: mercury monitor

    b. visible: colorimetric tape sampler

    c. infrared: MIRAN-type infrared analyzer

2. The matched filaments are typically replaced as a pair because the imbalance in resistance indicates the concentration of combustible gas. If a significant imbalance exists due to differences in the resistors themselves, then good measurements cannot be obtained.

3.
$$(0.25)\,(0.8\%) = 0.2\% \text{ or } \frac{2}{1,000}$$
$$\text{Concentration (ppm)} = \frac{2}{1,000} = \frac{2,000}{10^6} = 2,000 \text{ ppm}$$

4. Being able to measure 2,000 ppm or a somewhat lower value would be nearly meaningless in terms of being able to measure low concentration levels (e.g., those <500 ppm).

5. Silicone vapors can poison the catalytic activity of a platinum filament used in a combustible gas indicator, making it impossible to obtain measurements.

6. b. Because only small, grab samples are taken, results from detector-tube sampling make interpretation regarding worker exposures difficult.

7.
$$\text{TWA} = \frac{(15)(220\text{ppm}) + (15)(45) + (60)(30) + 390(17)}{(480)} = 25 \text{ ppm}$$

The measured TWA is one half of ACGIH TLV as a TWA (50 ppm).

8.

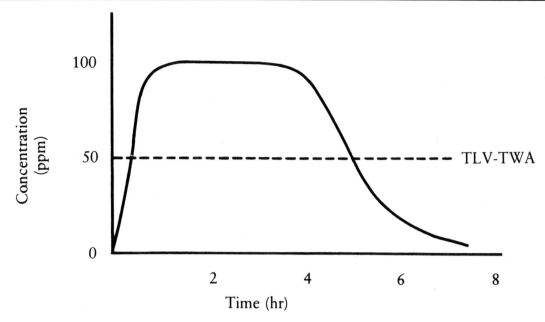

The operator is clearly overexposed with respect to the TLV-TWA.

9. Electrochemical polarographic cell

10. Carbon dioxide; devices absorb carbon dioxide into soda lime. The difference in air volume through soda lime absorber serves as an indicator of carbon dioxide concentrations.

11. See text, pp. 446–448

12. b. These two types of devices have entirely different principles of operation. Hence, an interference for one type should not affect the second device in an analogous manner.

Chapter 20—Answers

1. (a) Wet work, (b) isolation, (c) process enclosure.

2. Yes—This describes the very conditions when general exhaust ventilation should be used.

3. Substitution, isolation, process modification, enclosure

4. Evaluation of the effectiveness of installed devices

5. b

6. a. Biological monitoring reflects integrated exposures via all physiological routes of absorption; if media, such as blood and urine, continue to be contaminated, the controls are not working or the worker is being exposed to contaminants outside of the workplace.

7. a–d

8. From among these choices the student should choose methyl chloroform (1,1,1-trichloroethane) based upon pertinent industrial toxicology references.

9. The area sampling indicates concentrations close to the TLV. This sample may not be truly representative, and concentrations in this area may at times exceed the TLV. An industrial hygienist should recommend local exhaust ventilation, unless a better control (e.g., elimination, modification, or substitution) is available.

10. Methyl chloroform is the solvent of choice based on the alternatives listed. Personal protective equipment would need to match potential respiratory, ocular, and dermal routes of exposure. For most scenarios, suitable gloves, aprons, face shield, and approved organic respirator would be adequate. Certain exposure conditions might necessitate a supplied air respirator and more extensive skin protection.

11. Toxic decomposition products, such as hydrogen chloride, chlorine, phosgene; many others may result from contact of ultraviolet radiation with solvent vapors.

Chapter 21—Answers

1. c

2. a

3. d

4. b

5. b

6. c

7. d

8. b

9. a and c

10. a

11. a–d

12. b

13.

Contaminant	TWA Concentration (Range in mg/m³)	Applicability of TLV for mixtures
Total alkanes	2.3–267.1	None applicable
Trichloroethylene	4.0–11.9	None applicable
Toluene	0.3–0.7	Mix A
Butyl acetate	1.0–4.2	Mix B
Xylene	1.8–10.3	Mix A
Methyl amyl ketone	0.7–2.9	None applicable
Cellosolve acetate	3.3–23.6	Mix B

Comparison with the appropriate TLVs can be made by consulting Appendix B-1 in text, pp. 769–804, or the most recent edition of the ACGIH TLV and BEI Book; substances for which a TLV for mixtures might apply are noted.

14. a. Increased workload—Rapid hoisting of articles can make the device less effective, since the distances where capture is necessary are lengthened.

    b. Shortage of supply air—Supply is an essential component of these systems; if less than adequate, their performance degrades.

    c. Change in a paint component—This won't change the effectiveness per se but it may make the device chosen inappropriate.

    d. Increased forklift traffic—This will likely create crosscurrents that will make the device less effective.

15. adsorption

Chapter 22—Answers

1. Room volume = 150 ft × 60 ft × 10 ft = 90,000 cu ft
   To provide 6 air changes per hour:

2.
$$6 \times 90{,}000 \text{ cu ft} = 540{,}000 \text{ cu ft}$$
$$540{,}000 \text{ cu ft/hr} = 9{,}000 \text{ cu ft/min}$$
$$\frac{(9{,}000 \text{ cu ft/min})}{9 \text{ sq ft}} = 1{,}000 \text{ ft/min}$$

3. a

4. b

5. a. general office > b. labs and photo darkroom > c. industrial acid cleaning areas

6. See pp. 514–518 in text.

7. Consult p. 518 in text.

Chapter 23—Answers

1. c

2. Administration, knowledge of respiratory hazards and their assessment, control of respiratory hazards, selection of proper respiratory protective equipment, training, inspection, medical surveillance, and requirements as listed on text p. 523 and in the Appendices.

3. Charged metal fumes are attracted to the static-charged metallic compound; this improves the efficiency of the cartridge.

4. The high-temperature alarm obviously warns of an air temperature that would be dangerous; it also provides an indirect indication that something is wrong with the compressor, which could lead to air contamination (e.g., oil blow-by, carbon monoxide, other problems associated with deteriorating seals).

5. b

6. The opinion should reflect the fact that both the respirable (quartz) silica and nuisance dust concentrations are well within the current TLVs. Good industrial hygiene practice dictates that approved devices should be used when available.

7. The exposure approximates the TLV. Therefore, some form of respiratory protection is recommended. An organic vapor cartridge-type respirator should be employed.

8. This question indicates the limitations of air-purifying devices. Very likely an airline-type (air-supplying) device is needed.

9. a

10. b

11. b

12. eye irritation

13. Gas masks should not be used for any hazardous entry situation, because their adsorbent capacity may be inadequate for the length of the work task. However, such devices may be very useful for emergency egress.

14. No, the chemical composition of the adsorbent bed is identical; the devices differ with respect to the amount of adsorbent present. Like other air-purifying devices they can **never** be used in an oxygen-deficient atmosphere.

15.   Concentration = $\dfrac{0.003 \text{ mmHg}}{760 \text{ mmHg}} = (10^6) = 3.95$ ppm
           (worst case—ppm)

           Concentration = 3.95 ppm $(224.1/24.5) = 36.1 \text{ mg/m}^3$
           (worst case—mg/m³)

16.   Given that the OSHA PEL is 0.1 mg/m³ and that it is possible air concentrations could exceed 36.1 mg/m³, the respirator must furnish a protection factor of 361 (36.1/0.1). Erring on the side of caution, this would limit respirators which could be recommended to those having protection factors of 1,000 or greater.

           Any air-supplying devices would need to incorporate a full facepiece and would include airline devices when used in the pressure-demand or continuous-flow modes or possibly self-contained breathing air. An impervious suit would also be required.
           Heat stress problems associated with the use of an impervious suit should also be considered. Finally, the degree of protection necessary to safely carry out this task may suggest that the task needs to be conducted in a different manner so that human exposures are precluded.

17.   The respirator is not furnishing protection; from the information given the protective suit is providing dermal protection.

18.   Dermal contact is significant; the suit is not adequate but the respirator appears to be satisfactory.

19.   Biological monitoring indicates protection is inadequate; neither respiratory nor whole body protection is being provided.

(Please note: There are no specific answers to questions on Chapters 24, 25, 27, 28, 29, and 31.)

Chapter 26—Answers

(No specific answers are provided for questions 1–5.)

6.   Measures including FEF_{25-75}, FEV_1, and FVC can be useful. The specific application of these tests is not outlined, only their general availability.

7. Biological monitoring addresses integrated exposures, e.g., those from any physiological route of exposure (inhalation, percutaneous, ingestion) and through any means (environmental, occupational, behavioral). For a substance like fluoride, which is likely contacted through a variety of means, biological monitoring serves as a check on the effectiveness of air monitoring in the workplace or any other specific tools.

8. Doing biological monitoring without references values is a poor idea. Whether for an individual worker or a group of workers, it is necessary to compare results with a "benchmark" or reference values. Although research is sometimes done to establish what sort of reference value is reasonable, biological monitoring should not be performed unless well-documented reference values are available.

(Please note: There are no specific answers to questions on Chapters 27, 28, and 29.)

Chapter 30—Answers

1. The action level is generally one half of the permissible exposure level (see Appendix F, Glossary in text); as such, it represents an important criterion for decision-making. When sampling reveals concentrations above the action level, then action is required to do further sampling and to ameliorate the conditions that are causing these concentrations (or their counterparts in the case of physical, biological, or mechanical agents). Reference to the action level can facilitate effective discussions with plant management regarding priorities of occupational health and safety services.

2. You have 15 working days from the receipt of the citation.

3. See p. 680 in text. Training includes explanation of the requirements of the standard itself, identification of hazardous workplace operations, knowledge of the methods to detect hazardous chemicals, warning regarding unlabeled pipes, and others, as cited.

4. No, only those who are potentially exposed to lead. In this case air monitoring may provide a guide to those employees who should be screened (e.g., air exposures $> 30 \ \mu g/m^3$). Professional judgment should dictate employee placement in the blood lead screening program. Employees requesting to be tested should also be counselled.

5. b

6. a

7. d

8. b

9. a

10. Furnishing the employee a place of employment free from recognized hazards . . .; employer shall comply with occupational safety and health standards under the Act . . . the employee shall comply with occupational safety and health standards . . . pursuant to the Act.

(Please note: There are no specific answers to questions on Chapter 31.)

Community Safety and Health Membership—What It Means...

- It means keeping abreast of safety and health issues and developments through NSC materials.

- It means the opportunity to get together with other concerned individuals and groups, both volunteer and professional, in solving cooperatively the safety and health issues of the day;

- It means the opportunity to join the fight to drastically reduce the annual toll of 32,000 needless American deaths—a figure three times that of occupational fatalities.

There are two types of Community Safety and Health memberships:

INDIVIDUAL
For those persons having general interest in non-occupational, off-the-job, safety and health $42.

ORGANIZATIONAL
For not-for-profit, public service organizations such as civic, fraternal and other local groups interested in general community safety and health $53.

*Please note: you can still participate in a community safety and health division (see back page for list of divisions) if you are a member of the NSC through your corporation, organization or association.

MEMBERSHIP MATERIALS

All Community Safety & Health members will receive:

Family Safety & Health Magazine, quarterly.
Volunteer's Voice for Community Safety & Health Newsletter, bi-monthly.
Catalogs, Poster Directory, as issued.
Annual Report

MEMBERSHIP SERVICES AND PRIVILEGES

MEMBER/NON-MEMBER PRICES—Only National Safety Council members are eligible to take advantage of our pricing policy. Community Safety and Health members are entitled to purchase single quantities only!

LIBRARY—The Council Library is the largest of its kind in the world. A computer is utilized to retrieve data for staff responses to member requests for general and technical information.

VOTING/PARTICIPATION—Individual and Organizational members are entitled to one vote upon payment of annual dues.

MEMBER ASSISTANCE—The Membership Department will be glad to provide assistance in answering questions about membership, materials and services. They can offer guidance to help your accident prevention efforts. A representative is assigned to each account. Please write or call if you need assistance of any kind.

NATIONAL SAFETY CONGRESS—Annual event in which safety exhibitors and speakers are brought together for a week-long gathering.

JOIN THE TEAM!

Members are encouraged to become active in one or more of the Divisions listed below. No additional membership dues are required. If interested in taking part, please check one or more of the following boxes and we will send you additional information.

☐ **Home Safety**   ☐ **Youth Activities**

☐ **Educational Resources**   ☐ **Public Safety**

☐ **Community Service**   ☐ **Religious Leaders**

☐ **Agricultural Safety**

SIGN UP TODAY!

Eligibility for Community Safety and Health Membership

This National Safety Council membership is for those individuals, community and civic organizations, fraternal groups and others having an active interest in safety or health and, at the least, serving as an information service for keeping abreast of community safety and health interests. Other types of memberships are available to private and governmental organizations concerned with safety and health programs for employees. Accordingly, individual businessmen, commercial profit making groups and national or statewide organizations are not eligible for Community Safety and Health Membership.

Organization Name:_____

Individual Contact Name:_____ Contact Title:_____

Street Address:_____ City & State:_____

Telephone: (_____)_____ Annual Dues   Individual ☐ $42   Organization ☐ $53

Signature_____ Date_____

Please enclose first year's membership dues payable to:

National Safety Council 1121 Spring Lake Drive, Itasca, Illinois 60143-3201

A NON-GOVERNMENTAL, NOT FOR PROFIT, PUBLIC SERVICE ORGANIZATION   (708) 285-1121